Geomorphology and Bank Erosion of the Matanuska River, Southcentral Alaska

By Janet H. Curran and Monica L. McTeague

Prepared in cooperation with the Matanuska-Susitna Borough

Scientific Investigations Report 2011–5214

U.S. Department of the Interior
U.S. Geological Survey

U.S. Department of the Interior
KEN SALAZAR, Secretary

U.S. Geological Survey
Marcia K. McNutt, Director

U.S. Geological Survey, Reston, Virginia: 2011

For more information on the USGS—the Federal source for science about the Earth, its natural and living resources, natural hazards, and the environment, visit http://www.usgs.gov or call 1–888–ASK–USGS.

For an overview of USGS information products, including maps, imagery, and publications, visit http://www.usgs.gov/pubprod

To order this and other USGS information products, visit http://store.usgs.gov

Suggested citation:
Curran, J.H., and McTeague, M.L., 2011, Geomorphology and bank erosion of the Matanuska River, southcentral Alaska: U.S. Geological Survey Scientific Investigations Report 2011–5214, 52 p.

Contents

Contents—Continued

Figures

Figures—Continued

Tables

Conversion Factors and Datums

Conversion Factors

Multiply	By	To obtain
Length		
inch (in.)	2.54	centimeter (cm)
inch (in.)	25.4	millimeter (mm)
foot (ft)	0.3048	meter (m)
mile (mi)	1.609	kilometer (km)
Area		
square mile (mi^2)	2.590	square kilometer (km^2)
acre	4,047	square meter (m^2)
Flow rate		
foot per second (ft/s)	0.3048	meter per second (m/s)
cubic foot per second (ft^3/s)	0.02832	cubic meter per second (m^3/s)

Temperature in degrees Fahrenheit (°F) may be converted to degrees Celsius (°C) as follows:

$$°C=(°F-32)/1.8$$

Specific conductance is given in microsiemens per centimeter at 25 degrees Celsius (µS/cm at 25°C).

Concentrations of chemical constituents in water are given either in milligrams per liter (mg/L) or micrograms per liter (µg/L).

Datums

Vertical coordinate information is referenced to the North American Vertical Datum of 1929 (NAVD 29).

Horizontal coordinate information is referenced to the North American Datum of 1983 (NAD 83).

Altitude, as used in this report, refers to distance above the vertical datum.

Geomorphology and Bank Erosion of the Matanuska River, Southcentral Alaska

By Janet H. Curran and Monica L. McTeague

Abstract

Bank erosion along the Matanuska River, a braided, glacial river in southcentral Alaska, has damaged or threatened houses, roadways, and public facilities for decades. Mapping of river geomorphology and bank characteristics for a 65-mile study area from the Matanuska Glacier to the river mouth provided erodibility information that was assessed along with 1949–2006 erosion to establish erosion hazard data. Braid plain margins were delineated from 1949, 1962, and 2006 orthophotographs to provide detailed measurements of erosion. Bank material and height and geomorphic features within the Matanuska River valley (primarily terraces and tributary fans) were mapped in a Geographic Information System (GIS) from orthophotographs and field observations to provide categories of erodibility and extent of the erodible corridor. The braid plain expanded 861 acres between 1949 and 2006. Erosion in the highest category ranged from 225 to 1,043 feet at reaches of bank an average of 0.5 mile long, affecting 8 percent of the banks but accounting for 64 percent of the erosion. Correlation of erosion to measurable predictor variables was limited to bank height and material. Streamflow statistics, such as peak streamflow or mean annual streamflow, were not clearly linked to erosion, which can occur during the prolonged period of summer high flows where channels are adjacent to an erodible braid plain margin. The historical braid plain, which includes vegetated braid plain bars and islands and active channels, was identified as the greatest riverine hazard area on the basis of its historical occupation. In 2006, the historical braid plain was an average of 15 years old, as determined from the estimated age of vegetation visible in orthophotographs. Bank erosion hazards at the braid plain margins can be mapped by combining bank material, bank height, and geomorphology data. Bedrock bluffs at least 10 feet high (31 percent of the braid plain margins) present no erosion hazard. At unconsolidated banks (63 percent of the braid plain margins), erosion hazards are great and the distinction in hazards between banks of varying height or geomorphology is slight.

Introduction

The Matanuska River is a braided, glacial river that drains a 2,100-mi^2 basin in the Chugach and Talkeetna Mountains in southcentral Alaska (fig. 1). For 85 percent of the river length, multiple channels simultaneously convey streamflow across a wide area referred to as the braid plain. Unlike a flood plain, which is typically covered by shallow water during flooding that returns to the main channel once floodwaters recede, a braid plain can become occupied by a main channel at modest flow levels. When a channel erodes the bank at the edge of the braid plain, the braid plain expands.

Braiding on the Matanuska River is active, and most non-bedrock banks offer little resistance to erosion. For decades, the river has eroded private properties, a major regional highway, and smaller roads in southcentral Alaska. In the lowlands downstream of Palmer, where farmlands bordered much of the river, this movement of the channel resulted in periodic loss of farmland at least as early as 1956 (U.S. Army Corps of Engineers, 2003). By the 1990s, when individual homes and subdivisions had begun to replace agricultural land uses in this area, the increasing consequence of riverbank erosion led to erosion studies and site-specific erosion control efforts.

Previous investigations of erosion along the Matanuska River have focused on the area near Palmer. The most comprehensive recent studies include a historical erosion assessment using aerial photographs (U.S. Army Corps of Engineers, 2003) and an analysis of river processes and erosion control options with a focus on potential gravel mining (U.S. Department of Agriculture, Natural Resources Conservation Service, 2004). Structural erosion control near Palmer and farther upstream has been designed in response to the erosion of a particular area and includes dikes, riprapped toe protection, and various unengineered rock features.

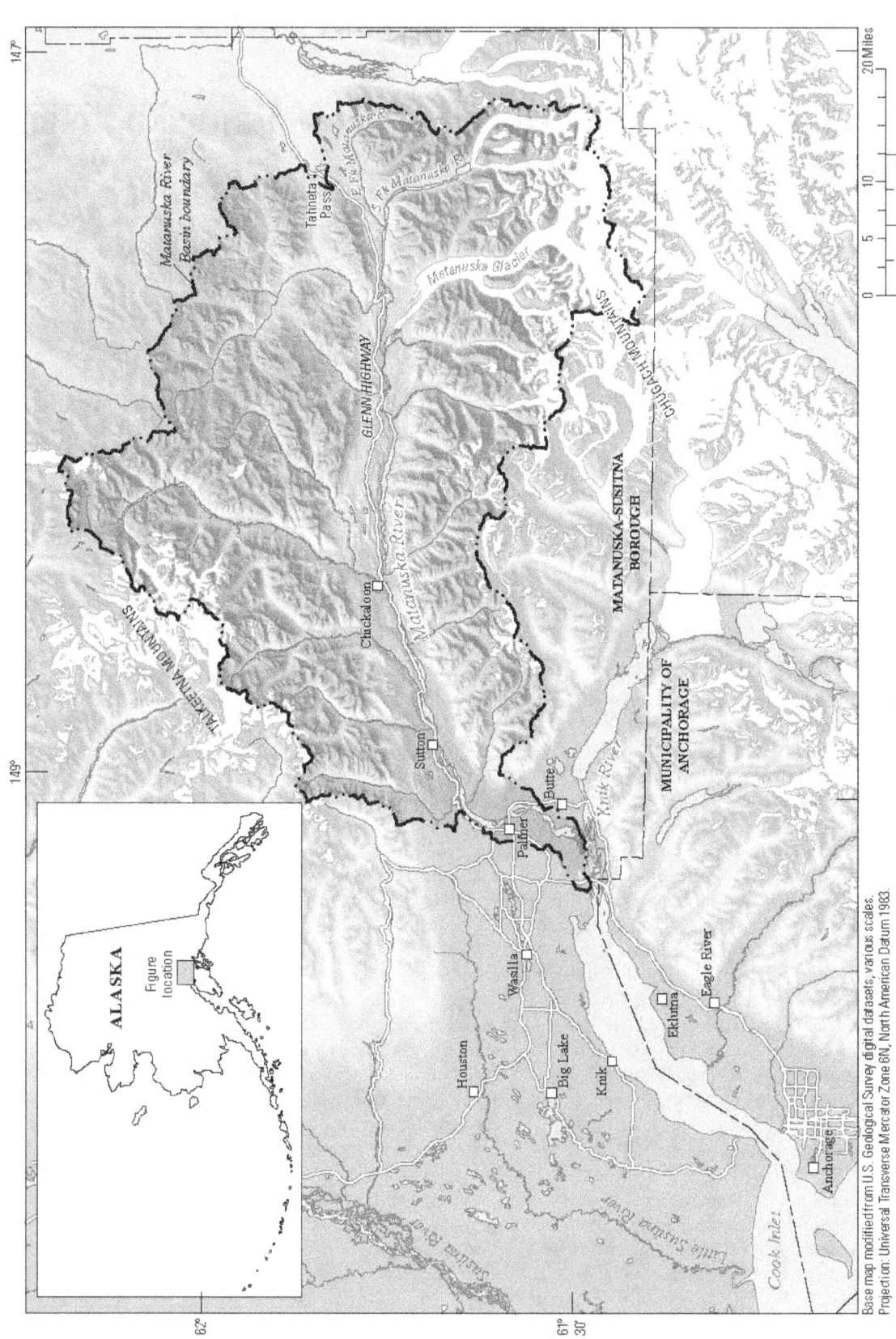

Base map modified from U.S. Geological Survey digital datasets, various scales.
Projection: Universal Transverse Mercator Zone 6N, North American Datum 1983

Figure 1.   Location of Matanuska River and vicinity, southcentral Alaska.

Challenges associated with erosion control, including expense, lack of a coordinated effort or responsible agency, and erosion of installed structural features, have prompted the Matanuska-Susitna (Mat-Su) Borough to consider nonstructural solutions. Development along the Matanuska River is being managed in the same way as other water bodies in the Mat-Su Borough, where structures can be placed within 75 ft of the shoreline and many property owners are unaware of the erosion hazards associated with the river. The Matanuska River Management Plan adopted by the Mat-Su Borough in 2010 (Mat-Su Borough, 2010) outlined goals embracing a mix of solutions, ranging from identifying where structural erosion control is required to erosion hazard education and development of regulations for high risk erosion areas. The USGS, in cooperation with the Mat-Su Borough, investigated geomorphology and bank erosion along the Matanuska River, from the Matanuska Glacier to the river mouth (fig. 2), to improve understanding of river processes

and develop tools for identifying bank erosion hazards. The study approach combined measurement of historical erosion with mapping bank characteristics and geomorphic features.

Purpose and Scope

This report presents new bank erosion and river process data for the Matanuska River and discusses bank erosion hazards in a geomorphic context. The study area encompasses most of the river to generate a comprehensive view of the frequency and distribution of erosion that can be used by future investigations of problem areas and solutions. Although erosion at specific areas became newsworthy during this study and various erosion control proposals emerged, this study does not address a particular erosion issue or location, nor does it address specific solutions. The intent of this report is to provide a tool and guidelines for application, rather than to definitively assign hazard categories.

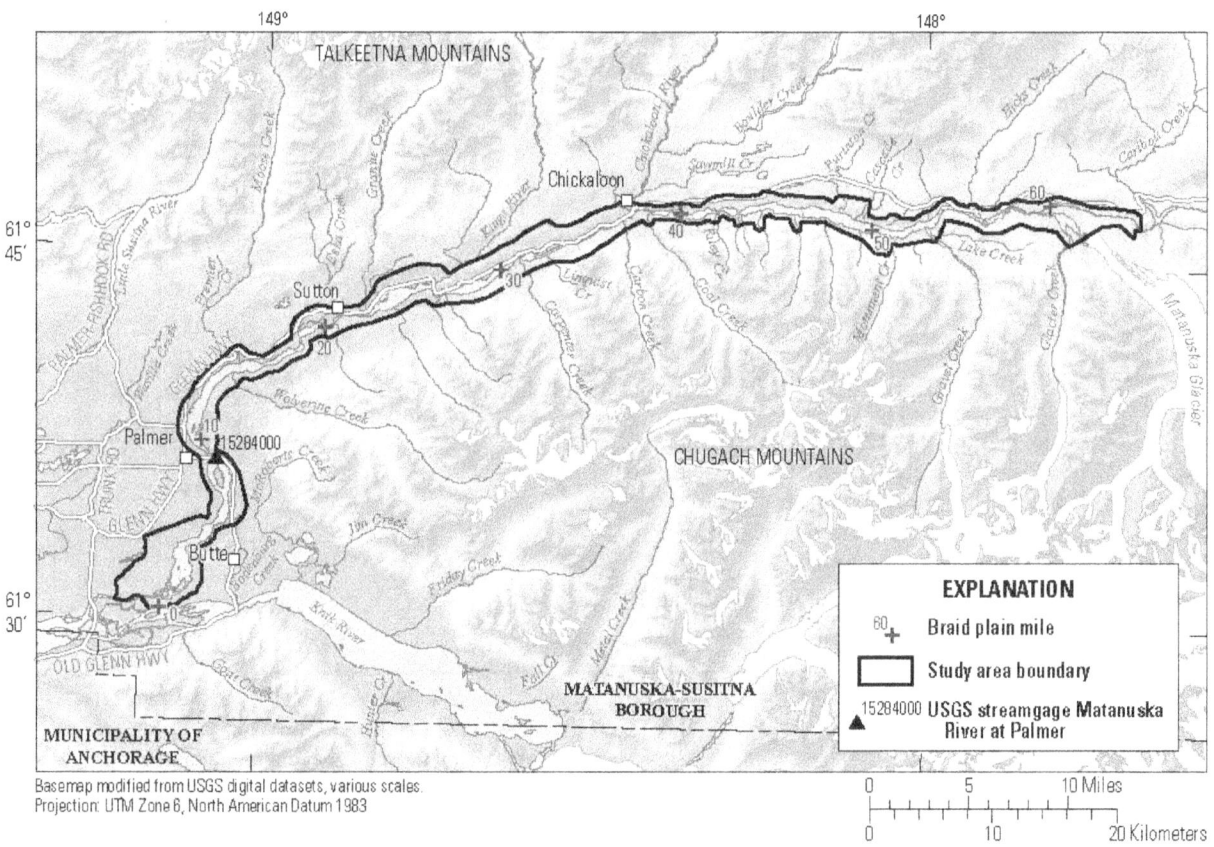

Figure 2.   Location of study area along the Matanuska River, southcentral Alaska.

Description of Study Area

Physical and Cultural Setting

The Matanuska River flows between the Chugach and Talkeetna Mountains through a mountain valley and across a lowland collectively known as the Matanuska Valley. The broad mountain valley contains small communities, including Sutton and Chickaloon, clustered around the Glenn Highway, a major regional highway that parallels the Matanuska River. The lowland contains the city of Palmer, which had a 2009 population of 8,428 (U.S. Census Bureau, 2009), and the smaller community of Butte. Outside the Matanuska River braid plain, rapid development of the Mat-Su Borough over the past several decades has resulted in sparse but increasing residential structures and public facilities near the river banks. Although parts of the braid plain are privately owned, few structures have been constructed there and activities generally consist of recreation. Limited commercial gravel mining has occurred near Palmer, and elsewhere, minor amounts of gravel have been mined for temporary construction projects.

The Matanuska Valley is a northeast-trending structural trough in rocks of Mesozoic and Tertiary age (Winkler, 1992; Wilson and others, 2009) that was occupied during the Pleistocene period by the Matanuska Glacier, whose present-day terminus is 50 mi east of Palmer. Glacial deposits can be identified near the present Matanuska Glacier, and deposits near Palmer have been interpreted as till, morainal, and outwash deposits in sequences up to a few hundred feet thick (Trainer, 1960; Freethey and Scully, 1980; Reger and Updike, 1983). Strong winds that entrain sediment from the braid plain or banks have left a prominent loess cap on many banks near the river (Trainer, 1961; Muhs and others, 2004) and modern sand dunes at sediment bluff tops near Palmer. New evidence and reinterpretation of the sediments and features in the Matanuska Valley suggest that transbasin flooding from paleo-glacial lakes in the Copper River Basin may have swept down the Matanuska Valley (Wiedmer and others, 2010). Flooding on this scale could have removed or reworked glacial sediments along the valley and resulted in the deposition of alluvial sediments.

The Matanuska River watershed descends from a maximum elevation of 13,000 ft in the Chugach Mountains to near sea level at the confluence with the Knik River. The mainstem of the river begins at the confluence of the South Fork Matanuska River, which drains several glaciers in the Chugach Mountains, and the East Fork Matanuska River, which drains the Tahneta Pass area at the low divide with the Copper River Basin. The prominent Matanuska Glacier, which stretches more than 20 mi through the Chugach Mountains and projects into the Matanuska Valley, discharges to the Matanuska River at numerous locations 8 mi downstream of this confluence. Other tributaries, the largest of which are Kings and Chickaloon Rivers, originate in the Chugach and Talkeetna Mountains and join the Matanuska River from upstream of the glacier to Palmer.

The Matanuska Valley has a maritime climate influenced by Gulf of Alaska weather systems that are moderated by the orographic effect of the Chugach Mountains, which creates a rainshadow with reduced precipitation. Mean annual precipitation is low along the valley bottom at Palmer and Sheep Mountain airport near the Matanuska Glacier (15 and 13 in., respectively) (Western Regional Climate Center, 2010), and increases to more than 80 in. at high elevations in the mountainous areas (Western Regional Climate Center, 2000). Temperatures are moderated by the maritime influence and by strong winds down the Matanuska Valley (U.S. Department of Agriculture, Natural Resources Conservation Service, 1998). At Palmer, the coldest month is January with an average low temperature of about 5°F and the warmest month is July with an average high temperature of about 67°F.

Geomorphic and Hydrologic Setting

The Matanuska River braid plain occupies a part of the Matanuska Valley. Upstream of Palmer, the braid plain is flanked by bedrock banks, glacial deposits, narrow fluvial terraces, and tributary fans in a confined valley; near and downstream of Palmer, the braid plain is flanked by broad glacial terraces in an unconfined valley. The river transports a lithologically diverse suite of sediment derived from the sedimentary and igneous rocks of the Talkeetna Mountains as well as metasedimentary and mélange-like accretionary rocks of the Chugach Mountains (Barnes, 1962; Winkler, 1992; Reger and others, 1996).

In addition to turbid mainstem channels, the braid plain contains clearwater side channels, which are shallow streams originating at springs within the braid plain or at tributaries to the Matanuska River. Disconnected from the Matanuska River except at their downstream ends, these streams commonly occupy channels abandoned by the mainstem and can form branching networks across braid plain bars. Clearwater side channels form important spawning habitat for sockeye, chum, and Coho salmon (*Oncorhynchus nerka, keta, and kisutch*, respectively) (Anderson and Bromaghin, 2009; Curran and others, 2011.)

Streamflow records for the Matanuska River from the USGS streamgage at the Old Glenn Highway bridge at Palmer, braid plain mile (BPM) 8.9 (USGS Station 15284000, Matanuska River at Palmer, Alaska) (fig. 2) show seasonal patterns and diurnal flows typical of Alaskan glacial streams (see fig. 2 in Wiley and Curran, 2003). From 35 years of daily records between 1949 and 2009, an envelope curve of daily mean discharge of the Matanuska River (fig. 3) shows a general streamflow decline through the fall and winter months, an abrupt rise with snowmelt in April and May, then a more gradual rise with glacier melt in June and July. Monthly mean streamflow was highest in June, July, and August during the period of record. Hydrographs for individual years

mirror the general pattern of an annual prolonged summer high flow period shown in figure 3 but smaller precipitation driven fluctuations created more variability on a daily to weekly scale.

The mean annual flow of the Matanuska River averaged 3,880 ft³/s at Palmer during the period of record from 1949 to 2009 (U.S. Geological Survey, 2009). The mainstem of the river is turbid from spring through fall, when glacial runoff is greatest, and is relatively clear beneath an ice cover in winter. Glaciers occupy 12 percent of the river basin, and small lakes and ponds occupy less than 1 percent of the basin (Curran and others, 2003).

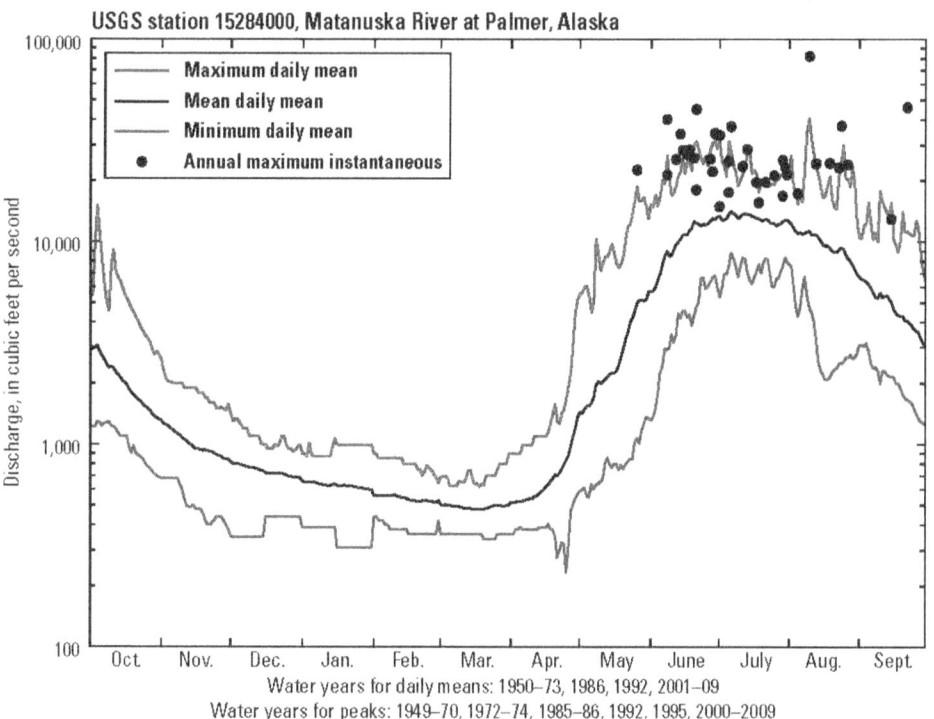

Figure 3.   Maximum, mean, and minimum daily mean discharge and annual maximum instantaneous discharge and calendar day of occurrence during the period of record for the Matanuska River at Palmer, Alaska, USGS gaging station 15284000.

Methods

Orthorectified aerial photographs and historical maps analyzed in a Geographic Information System (GIS) provided the basis for mapping banklines and geomorphic features in the 65-mi long study area, which included 138 mi of bank, and for assessing historical erosion in a subset of the study area with 133 mi of bank. Geomorphic characterization of the river also relied on visual field observations, field measurements, orthorectified modern and historical aerial photographs, oblique aerial photographs, and topographic maps.

Definitions

Braid Plain

The *braid plain* is defined as the area presently or formerly occupied by active channels of the river. For a consistent timeframe, this report refers to the *historical braid plain* as the area occupied by the river within the historical period. The *historical period* is loosely defined as about the 100 years preceding the year 2006. This includes the 57-year record of aerial photographs (2006–1949) plus the estimated minimum 40-year age of trees on the oldest channel-scarred surfaces in the 1949 photographs. The exact number of years in this period was less important to the study than the consistent, readily detectable relative age provided by channel scars. Isolated older patches without channel scars (containing vegetation estimated to be on the order of 200 years old) were included in the braid plain only where surrounded by younger, channel-scarred surfaces.

The braid plain contains the *active channel belt* (the suite of wetted channels and intervening *bars*, or gravelly surfaces with no vegetation visible on the orthophotographs) and vegetated surfaces categorized on the basis of their position as *braid plain bars* (vegetated braid plain outside the active channel belt) and *braid plain islands* (vegetated braid plain within the active channel belt) (fig. 4). Formerly part of the active channel belt, braid plain bars and islands can contain abandoned channels and bars and can remain disconnected from streamflow long enough for vegetation ranging from grasses and shrubs to mixed deciduous and spruce forest to become established before the surface is reincorporated into the active channel belt. Islands generally do not stand significantly higher than surrounding active channel belt. Isolated bedrock outcrops in the active channel belt stand up to tens of feet above the braid plain but are denoted separately as *braid plain bedrock features* rather than islands. Islands and braid plain bedrock features are relatively uncommon in the Matanuska River.

River miles are often used as a reference for location along a river, and are presented in appendix A (River_MI shapefile), but are less meaningful in a rapidly changing braided river. *Braid plain miles* (BPM) measured in miles along the center of the braid plain upstream of the river mouth are presented in appendix A (BraidPlain_MI shapefile) and used throughout the report. River and braid plain distances also are presented in kilometers (appendix A, River_KM and BraidPlain_KM).

Historically Flooded Area

Areas where evidence of recent flooding by the Matanuska River can be seen in aerial photography but conversion of the land to active channel did not occur were denoted as *historically flooded areas*. These areas contain light-colored areas and channels suggesting deposition of fresh sediment, and patches of dead trees interpreted to be inundation-killed. In a geomorphic sense, these areas are flood plains, defined as low-lying areas adjacent to a river that can be flooded by the river. However, these areas were not mapped following protocols for flood-plain management, and should not be interpreted as defining the extent of the flood plain.

Terrace

Terraces are defined as features occupied by the Matanuska River prior to the historical period. Flat-topped and higher than the braid plain, they are disconnected from the historical braid plain and typically receive no river streamflow, even during floods. No recent channel scars can be seen through terrace vegetation.

Upland

Upland is a general term used to refer to a valley bottom feature not formed by the modern Matanuska River or its tributaries. Uplands include bedrock, glacial deposits, and landslides, and can be part of the braid plain margin.

Data Sources

New true color orthophotography derived from 1:24,000 aerial photographs taken for this project on October 15, 2006, provided a shadow-free, leaf-off, and ice-free base for mapping banklines and geomorphic features. Additional orthophotography from 1949 and 1962 aerial photographs provided a historical comparison, and true color orthophotography from 2004 (downstream of Chickaloon) and satellite imagery from 2005 (upstream of Chickaloon) provided a historical comparison and coverage outside the mainstem corridor (table 1). Orthophotography for selected areas from additional dates (table 1) was used to confirm observations and allowed for additional analysis of bankline position. Historical maps published by the U.S. Surveyor General's Office that depict conditions from surveys in 1913–16 were orthorectified by USGS and used to compare bank positions near Palmer. Contact prints of the 1960 and 2006 photography were viewed using a stereoscope to assist delineation of banklines and geomorphic features. All orthophotography created for this project (table 1) was

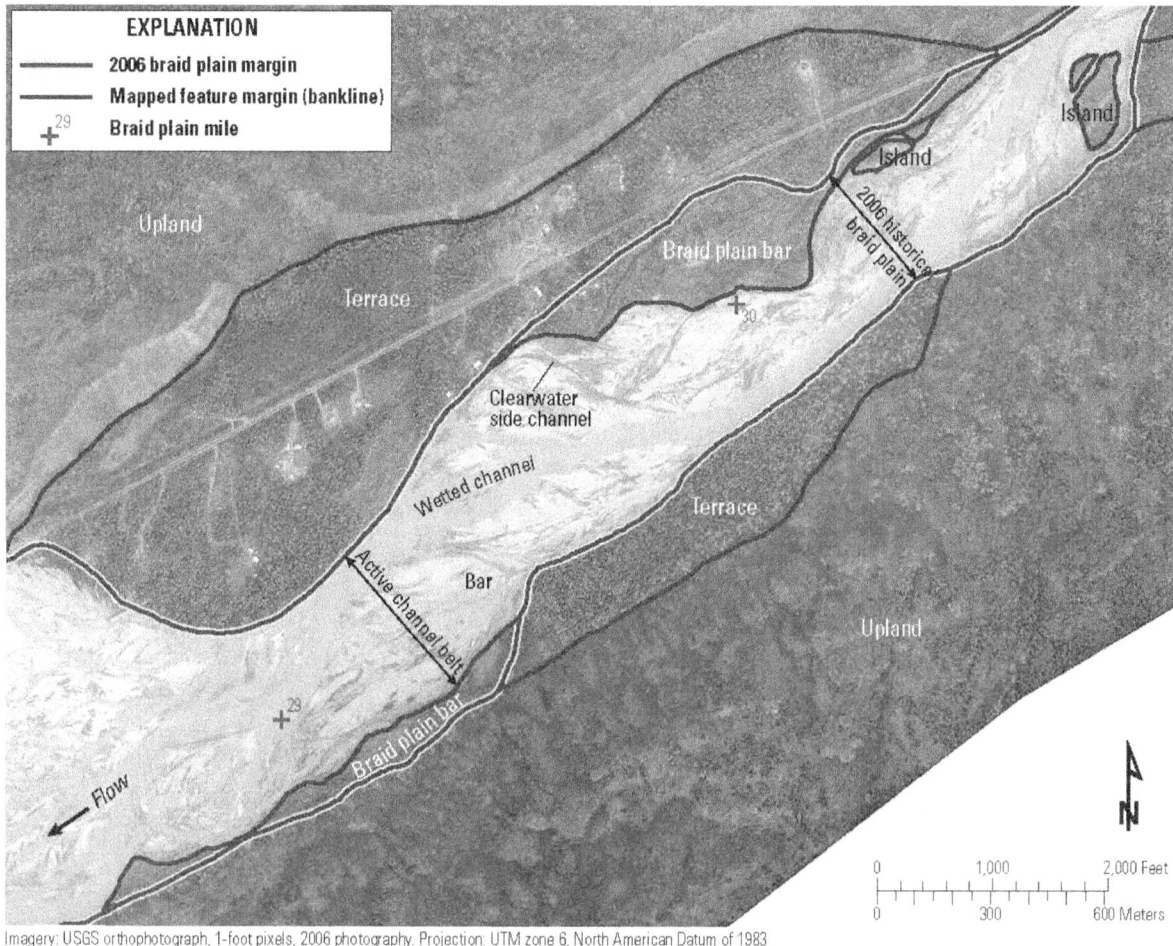

Figure 4. Typical valley bottom and braid plain features of the Matanuska River, southcentral Alaska.

orthorectified by a contractor and has a horizontal datum of North American Datum, 1983 (NAD 83). The projection for all GIS data associated with this report is in Universal Transverse Mercator (UTM) Zone 6, Meters.

Published and unpublished maps and digital datasets (table 1) generally were used to confirm observations and aid interpretation. Geologic maps (Winkler, 1992; Wilson and others, 2009), although coarse in scale, helped confirm locations of bedrock. Landslides shown in Detterman and others (1976) were transcribed directly into the banklines without field confirmation or further interpretation. Unpublished data (Kopczynski, Lehigh University, written commun., 2008) provoked further inspection of areas near Butte. Soils data (Clark and Kautz, 1998; Clark, 2006), although focused on the ground surface rather than river-level exposures, were reviewed in the early stages of mapping to help refine categories and selected line locations. Scanned versions of 1:63,360 USGS topographic maps provided elevation data for the longitudinal profile. High resolution topography acquired by Light Detection and Ranging (LIDAR) techniques in 2003 by others and converted to a useable datum for this project, as well as new 2006 data covering a subset of that area, enhanced detail and provided

an opportunity to compare riverbed elevations. Hillshade, elevational models, and slope breaks products created with ArcGIS® from the LIDAR datasets were used to identify features.

Although most mapping and analysis relied on orthophotography and GIS tools, short field campaigns ranging from driving and walking reconnaissance to raft supported reconnaissance and data collection from 2006 to 2010 provided field estimates of bank height, calibration of orthophotograph interpretations, ground photographs, and mapping refinements. Continuous tracks logged with a handheld Garmin® 76CSx Global Positioning System (GPS) documented field positions. GPS-Photolink™ software, a time-stamp coordinating software, linked field photographs to GPS coordinates. Photographs and video from a low-elevation aerial reconnaissance on August 14, 2007, provided detailed oblique imagery for closer inspection of various features.

Hydrologic data were obtained online from the USGS National Water Information System (NWIS) (http://waterdata. usgs.gov/ak/nwis/sw) for USGS streamgage 15284000, Matanuska River at Palmer. Selected streamflow statistics were obtained from USGS annual water data reports (U.S. Geological Survey, 2009).

Table 1.   Data sources used for mapping banklines and geomorphic features along the Matanuska River, southcentral Alaska.

[Media: LIDAR, light detection and ranging. **Discharge data** for USGS streamgage 15284000 from http://waterdata.usgs.gov/ak/nwis/sw. **Photograph type:** BW, black and white; CIR, color infrared. **Source:** MSB, Matanuska-Susitna Borough; NED, National Elevation Dataset; NRCS, U.S. Department of Agriculture, Natural Resources Conservation Service; USGS, U.S. Geological Survey. **Abbreviations:** ft³/s, cubic foot per second; ft, foot; n/a, not applicable; –, not available]

Media	Year	Month/Day	Matanuska River discharge at Palmer (ft³/s)	Photograph type	Source	Extent	Scale	Pixel resolution (ft)
Full study area								
Aerial orthophotography	1949	8/10, 8/14, 8/15	10,900–13,400	BW	USGS (this project)	River mouth to South Fork Matanuska River	1:50,000	2.0
Aerial orthophotography	1962	6/6 (1960)[1], 5/24 (1962)	7,840 (1960), 3,320 (1962)	BW	USGS (this project)	River mouth to South Fork Matanuska River	1:24,000 (1960[1]), 1:40,000 (1962)	2.0
Aerial orthophotography	2006	10/14	6,390	Color	USGS (this project)	River mouth to South Fork Matanuska River	1:24,000	1.0
Topographic map	various	n/a	n/a	n/a	USGS	Nearly statewide	1:63,360	n/a
Geologic map	1992	n/a	n/a	n/a	Winkler (1992)	Anchorage 1° × 3° Quadrangle	1:250,000	n/a
Digital elevation map	2008	n/a	n/a	n/a	USGS - NED	Statewide	n/a	197
Digital geologic map	2009	n/a	n/a	n/a	Wilson and others (2009)	Cook Inlet region	1:250,000	n/a
Part of study area								
Aerial orthophotography	1978	8/25	–	CIR	MSB	River mouth to Palmer	1:64,000	6.6
Aerial orthophotography	1985	9/21	4,700	Color	MSB	River mouth to Palmer	1:36,000	6.6
Aerial orthophotography	1990	7/19, 8/12	–	Color	USGS (this project)	River mouth to Palmer	1:40,000	2.0
Aerial orthophotography	1996	n/a	n/a	Color	MSB	River mouth to Palmer	1:24,000	6.6
Aerial orthophotography	2004	6/25, 6/26, 8/9, 8/10	23,900 (June), 10,800–11,100 (August)	Color	NRCS	River mouth to Chickaloon	1:24,000	3.3
Satellite orthoimagery	2005	8/11	17,000	Color	MSB	Chickaloon to Matanuska Glacier	n/a	3.3
Map	1915, 1916	n/a	n/a	n/a	U.S. Survey or General's Office	River mouth to Moose Creek (T16N, R1E; T17N, R1E; T17, R2E; T18N, R2E)	40 chains to an inch (1:31,680)	n/a
Geologic map	1976	n/a	n/a	n/a	Detterman and others (1976)	Moose Creek to Matanuska Glacier (Castle Mountain-Caribou fault system)	1:63,360	n/a
LIDAR	2003	November	n/a	n/a	USGS (this project) adjustment to NRCS (2004)	Circle View Subdivision to Old Glenn Highway bridge	Point spacing 5 ft	6.6
LIDAR	2006	October	n/a	n/a	USGS (this project)	Circle View Subdivision	Point spacing 2.8 ft	6.6
Soils map	1998	n/a	n/a	n/a	NRCS	Matanuska-Susitna Valley	1:24,000	n/a
Soils map	2006	n/a	n/a	n/a	NRCS	Chickaloon Village Lands	1:24,000	n/a
Unpublished digital geologic map	2008	n/a	n/a	n/a	Kopczynski (written commun., 2008)	River mouth to Palmer (Anchorage C-6 SW quadrangle)	1:25,000	n/a

[1]1960 photography was used to fill in the orthophotograph in a small area near the river mouth where 1962 photography was not available.

Mapping Banklines, Geomorphic Features, and Erosion Control Features

Lines defining the margins of braid plain and geomorphic features within the valley formed the basis for analysis of erosion and assessment of erosion hazards. Banklines were drawn in a Geographic Information System (GIS) and labeled with bank material and bank height information, as well as positional information including position with respect to the braid plain (Banklines_2006 shapefile). A geomorphic map (GeomorphicFeatures shapefile) and erosion control features map (ErosionControlFeatures shapefile) were prepared for 2006 conditions, and historical braid plain margins and braid plain features were mapped for 1949, 1962, and 2006 conditions (BraidPlainMargin_1949, BraidPlainMargin_1962, and BraidPlainMargin_2006, respectively). GIS shapefiles are presented in appendix A.

Bankline Delineation

Banklines were delineated on-screen by heads-up digitizing in a GIS using ArcGIS® at a scale of 1:2,300 for the post-2000, higher resolution orthophotography and up to 1:3,200 for older, lower resolution orthophotography. Toes of banks, rather than tops of banks, were mapped because tops were not consistently discernible from aerial photography and the change to the toe more adequately measured river related erosion on taller banks, also subject to other forms of erosion.

Banklines were ground verified with opportunistic and targeted reconnaissance efforts including driving, hiking, and boating surveys between June 2006 and August 2008. About 20 percent of the lines were delineated in the field. Reference hardcopy and digitized geology and soils maps of the Matanuska Valley and the lower Cook Inlet (Detterman and others, 1976; Winkler, 1992; Clark and Kautz, 1998; Clark, 2006; Kopczynski, Lehigh University, written commun., 2008) were used to help distinguish geomorphic features. Digital mapping of Cook Inlet (Wilson and others, 2009) that became available after most banklines were drawn was used to check and refine lines. Low elevation photographs from an aerial flight survey on August 14, 2007, were linked to tracks from a GPS to provide additional references for the banklines.

Banklines were drawn for fluvial features within the valley, including braid plain features, terraces, tributary fans, and areas outside the braid plain with visible evidence of historical flooding (here termed historically flooded areas). All vegetated surfaces within the braid plain likely to remain exposed at typical summer high flows (from fig. 3, the mean of the daily mean streamflows ranged from 10,000 to 14,000 ft³/s from mid-June to mid-August at the USGS gage in Palmer) were mapped as vegetated islands or braid plain bars. The 2004 orthophotography and 2005 satellite imagery were collected during typical summer high-flow conditions (table 1), thus providing a recent reference for defining the active channel belt and recategorizing a few vegetated surfaces

in the process of becoming fluvially reworked as active channel. The area mapped as active channel consists primarily of wetted channels and unvegetated bars but may include bars with small remnants of vegetation or likely to become inundated at typical peak flows. Banklines also were drawn for selected geomorphic features adjacent to the Matanuska River reflecting conditions pertinent to bank erosion or the geomorphic history of the river, including banks modified by roadway or railroad construction, mapped landslides, areas of bedrock likely to have been fluvially exposed, and areas influenced directly by the Matanuska Glacier. Sections of the abandoned railroad grade paralleling the Matanuska River from Palmer to Chickaloon that were visible in 1949 orthophotography were outlined in a separate file by digitizing on-screen (appendix A, Railroad shapefile).

Bank Characteristics

Bank erosion along most rivers can be related in varying degrees to bank erodibility. Bank material and bank height were the most readily determined bank erodibility factors for such a large study area on the Matanuska River, and were determined for all 2006 banklines within the valley and braid plain. Features mapped as landslides in Detterman and others (1976) were noted. A positional descriptor identified bank location relative to the braid plain margin (table 2). Definitions and categories of mapped bank characteristics are shown in table 2.

Bank material categories were designed to distinguish between easily erodible and relatively resistant banks, and to be easily applied from aerial photography over a large area. Bank materials were assigned to each line segment based on the material that was exposed at the base of the bank or at river level. In most cases, bank materials were visible or easily interpreted from the field and reference data sets. In areas where the bank was obscured by vegetation, surrounding materials or geomorphic indicators guided assumptions. Because of the importance of bedrock to bank erodibility, additional descriptors were added for bedrock integrity (solid versus blocky) and bedrock height. Where banks were vertically stratified with a bedrock strath, or bench, at the base of the bank overlain by unconsolidated sediments; the material was recorded as bedrock to reflect the erodibility of the material at river level, but given a bedrock height separate from overall bank height and an identifier to denote the presence of unconsolidated overburden (table 2). In areas where the bank material consisted of longitudinally intermittent sediment and bedrock outcrops not long enough to parse the line into segments, we applied an identifier to denote a combination of mixed unconsolidated sediment with bedrock exposures. Opportunistic visual estimation of bank-particle size classes (sand, gravel, cobbles, and boulders) documented at 65 locations spanning the study area guided interpretation of geomorphic origin and construction of bank material categories.

Table 2. Definitions of bankline attributes.

[Attribute category uses field name in GIS shapefile Banklines_2006 in appendix A. **Abbreviation:** ft, foot]

Attribute category and name	Definition
Material	Bank material
bedrock	Rock in its native position or rock dislodged by erosion or landsliding but not transported an appreciable distance by the river. Unconsolidated sediment overlies rock in thicknesses ranging from 0 to more than 100 ft.
unconsolidated sediment	Sediment consisting of particle sizes ranging from silt to boulders that is not visibly consolidated. Typically modern fluvial deposits or glacial outwash, also includes colluvium.
consolidated sediment	Sediment consisting of particle sizes ranging from silt to boulders that appears consolidated by virtue of its ability to stand in steep outcrops. Restricted to glacial till occurring near the Matanuska Glacier.
glacier	Glacial ice and ice-cored debris near the terminus of the Matanuska Glacier.
artificial fill	Fill material consisting of boulder-sized riprap or unsorted fill ranging from silt to boulders. Includes erosion control features, abandoned railroad fill, and roadway fill.
artificial line	A line drawn for convenience of analysis that has no physical meaning.
Height	Bank height
low	Height of bank above summer river level is less than 10 ft.
moderate	Height of bank above summer river level is 10–20 ft.
high	Height of bank above summer river level is greater than 20 ft.
bx_char	Bedrock character
solid	Bedrock that forms a continuous face along the river.
blocky	Bedrock that appears as knobs or as boulder-sized blocks dislogded from a larger outcrop but not transported an appreciable distance by the river.
intermittent	Bedrock that appears as discontinuous outcrops along the river, alternating with outcrops of unconsolidated sediment.
bx_height	Bedrock height
river level	Height of bedrock within bank is at river level (0 ft).
low	Height of bedrock within bank is less than 10 ft.
moderate	Height of bedrock within bank is 10–20 ft.
high	Height of bedrock within bank is greater than 20 ft.
Location	Lateral position of bank line with respect to braid plain
in braid plain	Bank is within the braid plain.
outside braid plain	Bank is outside the braid plain.
braid plain margin	Bank forms the edge of the braid plain. Considered the present river bank. Denoted as braid plain and upland margin when it also forms the edge of the upland.
fan margin	Bank forms the edge of a modern fan. Considered the present river bank and edge of the braid plain.
historically flooded area margin	Bank forms the edge of a historically flooded area, defined as an area where evidence of flooding was detected from aerial photography in 1949, 1962, or 2006. Denoted as historically flooded area and upland margin when it also forms the edge of the upland.
vert_strat	Vertical stratification of bank
stratified	Vertically stratified bank consisting of a lower unit of bedrock and an upper unit of unconsolidated sediment.
pub_slide	Published landslides
landslide	Area mapped as a landslide in Detterman and others (1976).

Bank height was most commonly visually estimated in the field. At selected banks, heights were measured using a tape, measured using a laser range finder equipped with angle measurement capabilities, or computed from LIDAR data to provide calibration of visual estimates. Other bank heights were estimated from aerial photography (viewed in stereo when necessary) or ground photography, or were extrapolated from adjacent banks where obscured. The summer river water level, which provided a convenient reference despite variations of several feet, was adopted as a datum for all bank heights regardless of bank location in the valley. Bank heights were grouped into three categories: low (less than 10 ft), moderate (10–20 ft), and high (greater than 20 ft).

Geomorphic Features

Geomorphic features were outlined by intersecting banklines for 2006 to form polygons. Minimum mappable size varied depending on source imagery resolution and distinctness of the feature, but the minimum size mapped consistently across the study area was 0.3 acre. Field mapped features too small for the level of detail in the imagery were combined with adjacent features. Although banklines indicate the location, height, and material present at the margin of uplands (such as bedrock bluffs, or glacial terraces), no effort was made to systematically close off and distinguish between upland types. As a result, the geomorphic features map in appendix A displays only features contained within the river valley and selected uplands relevant to the study.

Geomorphic feature attributes were assigned from selected bankline characteristics (table 3). Although the material of the bounding banklines generally was consistent and could be transferred to the polygon, bank height varied, requiring an interpretive assignment of dominant bank height for each polygon. Bank material categories were modified slightly from bankline categories for simplification.

Each feature was assigned to a major feature type (braid plain, fluvial deposits, modified land, or uplands), given an interpretive name (such as terrace or braid plain bar), and, where applicable, subdivided into categories reflecting feature age or position relative to the river. Feature identifiers and descriptions are listed in table 4 and in the GeomorphicFeatures shapefile in appendix A. Features were occasionally mapped with more detail than required to place them in age categories, such that a feature with two distinct vegetation (and thus age) patches might be divided into two adjacent features with identical interpretive names. These adjacent features can be treated as identical for the purposes of this report.

Table 3. Definitions of geomorphic feature attributes.

[Attribute category uses field name in GIS shapefile Geomorphic Features in appendix A. **Abbreviations:** ft, foot]

Attribute category and name	Definition
Material	Material
bedrock	Rock in its native position, usually isolated knobby outcrops. Only bedrock affected by the river was mapped as a geomorphic feature; refer to banklines for margins of bedrock uplands.
bedrock and sediment	Contains bedrock that appears as discontinuous outcrops along the river, alternating with outcrops of unconsolidated sediment.
unconsolidated sediment	Sediment consisting of particle sizes ranging from silt to boulders that is not visibly consolidated. Typically modern fluvial deposits or glacial outwash, also includes colluvium.
artificial fill	Fill material consisting of boulder-sized riprap or unsorted fill ranging from silt to boulders. Includes erosion control features, abandoned railroad fill, and roadway fill.
glacier	Matanuska Glacier.
landslide	Area mapped as a landslide by Detterman and others (1976).
Height	Feature height above river
low	Height of feature above summer river level is less than 10 ft.
moderate	Height of feature above summer river level is 10–20 ft.
high	Height of feature above summer river level is greater than 20 ft.

Table 4.   Definitions of geomorphic features mapped for 2006 conditions on the Matanuska River, southcentral Alaska.

[The historical period is defined by imagery from 1949 to 2006, extended by vegetation to a total of about 100 years. **Feature ID**: endings on the field are numeric for features with quantifiable ages (braid plain features) and are lower case for features where relative age is indicated by successive height and distance away from the river, but could not be correlated between locations. For example, all braid plain bars labeled BB3 were vegetated in 1949 and can be considered the same age class, but the age of the third oldest fan surface at one tributary (labeled Fc) was not necessarily the same age as the Fc fan surface at another tributary]

Feature name	Feature ID	Description
Braid plain feature types		Fluvial features formed by the Matanuska River or its tributaries within the historical record
active channels	AC	Wetted channels and unvegetated bars within the braid plain.
braid plain bar	BB1, BB2, BB3, BB4	Vegetated bar within the braid plain. May be adjacent to but not surrounded by active channel. Typically occurs as elongated strip adjacent to the braid plain margin. Numeric ending increases with age of feature:
braid plain fan	BF	Tributary fan deposited in the braid plain within the historical record.
braid plain island	BI1, BI2, BI3, BI4	Vegetated bar surrounded by active channels. Numeric ending follows convention for braid plain bar.
braid plain bedrock	BBX	Bedrock within the braid plain, occurring as isolated blocky outcrops ranging in size from megaboulders to 26 acres.
Fluvial deposit feature types		Fluvial features formed by the Matanuska River or its tributaries before the historical record
terrace	Ta, Tb, Tc	Surface occupied by the Matanuska River before the historical record. Typically extensive, flat, forested, elongated strips parallel to the river. Typically composed of fluvial deposits but may also be the result of fluvial erosion of bedrock or other materials.
tributary fan	Fa, Fb, Fc, Fd, Fe	Fans formed by tributaries to the Matanuska River. Typically fan-shaped in planform and slightly conical such that the outcrop along the Matanuska River was highest near the midpoint.
		Fa is that part of tributary fan that was active in the historical record. Includes fans occupied by the active tributary in 1949, 1962, or 2006, and fans vegetated in 1949 that contain channel scars and vegetation indicative of recent occupation.
		Fb-Fe: Various orders of fan surfaces, each older than the historical record and each increasing in age or height. May be paired across the modern fan. Fb may contain the active channel for streams that did not produce an active fan in the historical period.
historically flooded area	HFAa, HFAb	Low-lying area where evidence of flooding (open water, fresh sediment) is visible on orthoimagery but where active channels did not persist long enough to remove vegetation and convert the area to braid plain.
undifferentiated mainstem/fan deposit	U	Fluvial surface deposited or reworked by the Matanuska River and/or its tributaries. Specific processes not differentiated. May be gently sloping. May include evidence of flow parallel to modern channel. Positioned below or adjacent to older tributary fans. Extends beyond extent of typical tributary fan to form elongated apron. Interpreted as including fan sediments reworked by mainstem flow.
Modified land feature types		Anthropogenically created feature. Includes artificial fill, roads, and abandoned railroad bed
modified	OMD	Modified braid plain. Fill placed in braid plain, or mechanically disturbed braid plain surface.
roadway	ORD	Roadway. Fill embankments of the Glenn Highway.
railroad	ORR	Abandoned railroad bed constructed by 1917. Contains varying thicknesses of fill, generally including cobble-sized angular clasts.
Uplands feature types		Valley bottom features not formed by the modern Matanuska River or its tributaries. Mapped uplands include selected relevant features of limited size. Unmapped uplands are shown in the uplands shapefile and include bedrock, glacial outwash terraces, colluvium, and morainal material.

Table 4. Definitions of geomorphic features mapped for 2006 conditions on the Matanuska River, southcentral Alaska.—Continued

[The historical period is defined by imagery from 1949 to 2006, extended by vegetation to a total of about 100 years. **Feature ID**: endings on the field are numeric for features with quantifiable ages (braid plain features) and are lower case for features where relative age is indicated by successive height and distance away from the river, but could not be correlated between locations. For example, all braid plain bars labeled BB3 were vegetated in 1949 and can be considered the same age class, but the age of the third oldest fan surface at one tributary (labeled Fc) was not necessarily the same age as the Fc fan surface at another tributary]

Feature name	Feature ID	Description
fluvially affected bedrock	OBX	Bedrock that appears to have been fluvially stripped of overlying sediments, or bedrock surrounded by fluvial sediments.
glacier	OG	Matanuska Glacier.
inactive fan	OIF	Inactive fan. Gully heads and fan deposits that lack evidence of modern streams of sufficient size to generate the fans present. Occur along the edges of glacial terraces. Detail needed to map these features required LIDAR imagery, which was available only for a short reach downstream from Palmer.
landslide	OLS	Landslide mapped by Detterman and others (1976).
recent landslide	OLS recent	Landslide near braid plain margin with evidence of recent movement observed in field. Evidence of recent activity included seepage on bank face, slumps, leaning vegetation.
tributary valley	OTV	Tributary valley. Valley deposits along tributaries to the Matanuska River. Lack the fan shape and elevation of tributary fans.
undifferentiated glacial/ fluvial deposit	UGF	Surface deposited or reworked by the Matanuska Glacier and/or the Matanuska River. Likely morainal deposits reworked by river flow.

Erosion Control Features

Extensive or engineered erosion control structures and bank protection existing in 2006 were mapped to provide a baseline inventory and determine the extent of protected areas. This compilation of erosion control features updates a list in section 4.b of U.S. Army Corps of Engineers (2003) by providing features omitted from the list and providing feature locations in a GIS. All features visible from orthophotography or LIDAR were mapped and the locations of 91 percent of the features were confirmed with field observations. Less formal or less engineered features such as arrangements of car bodies, or tires or trees cabled to banks, were present in short reaches but their questionable efficacy and limited detectability from orthophotography precluded their inclusion in the inventory. Each erosion control feature was categorized by feature type (dike, bank armoring, levee) and concerned entity (highway, railroad, commercial, residential, or public facility), and given a location description. Location descriptions included the U.S. Army Corps of Engineers (2003) descriptor wherever applicable. Feature lengths computed from a line drawn along the top of the feature, regardless of orientation with respect to the river, formed a reasonable approximation of the length of bank protected.

Geomorphic Reaches

Reaches are considered a length of a river having similar form and fluvial process and provide a convenient basis for analysis and discussion. Initial aerial photograph inspection of braid plain width and extent of braiding, coupled with preliminary field observation of bank height, was used to define six reaches ranging in length from 6 to 20 braid plain miles (fig. 5). Informal names assigned to each reach denote local communities or tributaries. Reach 7, which was defined but not included in this study, extends from the Matanuska Glacier to the origin of the South Fork Matanuska River at Powell Glacier.

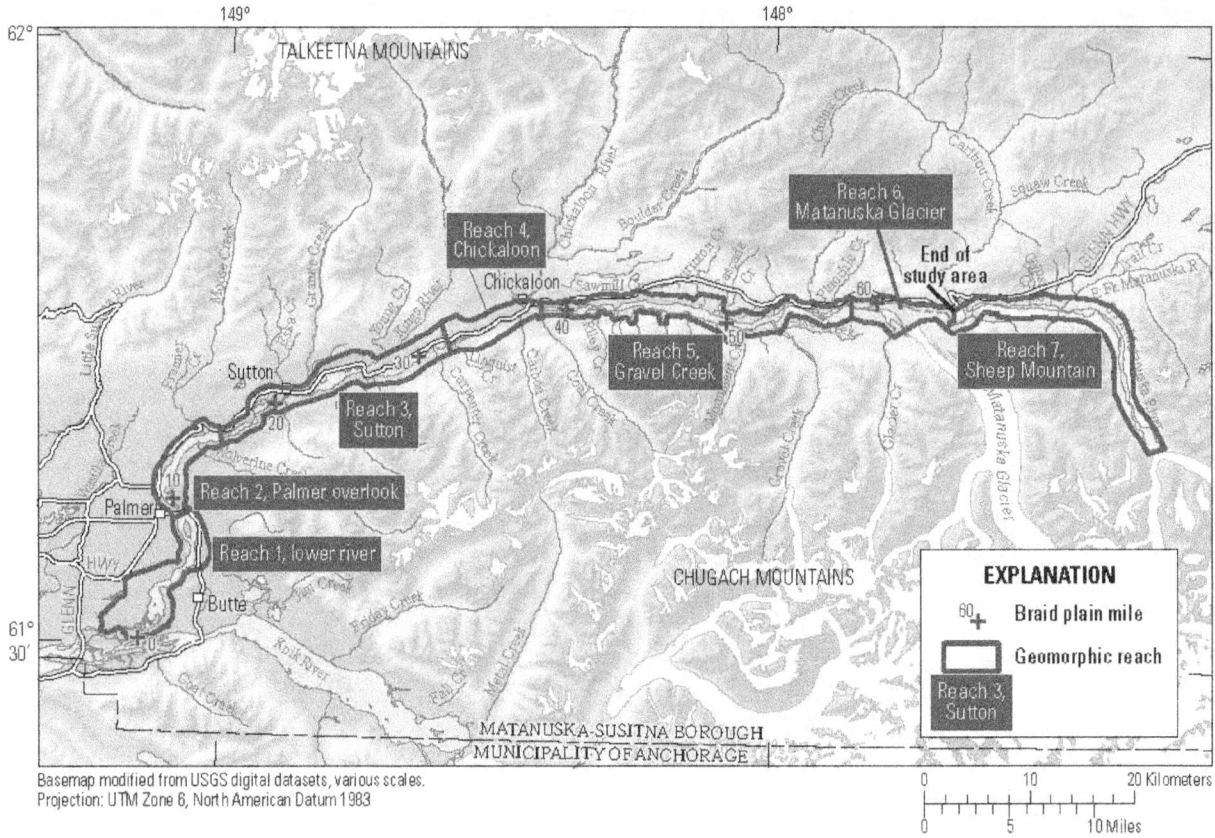

Figure 5. Geomorphic reaches of the Matanuska River, southcentral Alaska.

Geomorphic Measurements

Bed and Bank Material

Adequate measurement of bed and bank material throughout the study area would have required a substantial field campaign or remote sensing not necessary for this project and was substituted with visual classification to provide a more general depiction of grain size. Comments listing particle classes present (boulders, cobbles, gravels, sand) were noted at 65 bank locations, including braid plain margins and banks of channels within the braid plain, which represent bed material. Grain size measurements collected at the heads of mid-channel bars at BPM 49.0 (near Conglomerate Creek) and BPM 41.3 (near Riley Creek) provided data for determination of median grain size. The measurements were made using a gravelometer (metal template) to measure 100 particles on a rectangular grid spaced by one pace length of the observer and were collected near the water surface on June 26, 2008, a day during summer high flows with a mean streamflow of 8,360 ft³/s. Photographs documenting the bed or banks also provide a relative measure of grain size.

Longitudinal Profile and Slope

The longitudinal profile of the Matanuska River was plotted as the river length between known elevations. River length was determined from the digitized centerline through the largest channel in the 2006 orthophotograph (appendix A, RiverCenterline shapefile). Elevations were obtained by digitizing contours from USGS topographic maps at a 1:63,360 scale, the most detailed available for the whole study area. The contour interval was 100 ft, except below elevation 200 ft, where it was 50 ft. River slope was computed as total elevation decline divided by river length.

Alternate profiles using the available USGS 60 m (197 ft) Digital Elevation Model (DEM) and tracks from handheld GPS units were explored but were not as useful. Elevations from adjacent uplands influenced the DEM pixels, making detailed analysis inaccurate, and elevations from handheld GPS units were inaccurate. For the 2-mi long segment of LIDAR acquired for the lower river, elevations were extracted from 10 m (33 ft) transects.

Sinuosity

Sinuosity, defined as the ratio of river length to valley length, provides a measure of channel curvature. River length was determined from the digitized 2006 centerline, and braid plain length was determined from a braid plain centerline created by an automated GIS procedure that drew transects between braid plain boundaries and connected the midpoints of these transects (appendix A, BraidPlainCenterline shapefile). Values in this report represent 2006 conditions and were determined from the river length and braid plain of each geomorphic reach.

Braid Plain Age Estimates

Establishing the frequency of channel occupation at a particular location using repeat photography was not feasible given the mobility of the Matanuska River and the available photography. An alternative metric to provide a sense of the mobility of the channels was obtained by mapping the mosaic of various generations of vegetation present on braid plain

bars and islands. The area-weighted average age of the braid plain vegetation is an indication of how long, on average, surfaces remain stable before being reworked by the river. It cannot, however, predict the longevity of any one particular surface.

Vegetation colonization and succession patterns helped establish estimates of braid plain age, extending the historical record beyond the 1949 photography and aiding classification of braid plain surfaces. Estimated vegetation age based on cover density and canopy development was determined from a series of approximately decadal orthophotographs. Selected areas near Palmer that had orthophotography available for 1949, 1962, 1978, 1985, 1996, and 2006 were examined. Vegetation generally became visible a decade after the last surface disturbance and consisted of a patchy cover of herbaceous species and small shrubs, typically denser along margins of abandoned channels (fig. 6). After 2 to 3 decades, herbaceous and shrub cover fully covered the surface. Tree canopies appeared by 3 to 4 decades after the last disturbance, and after a minimum of 4 decades individual tree crowns became visible.

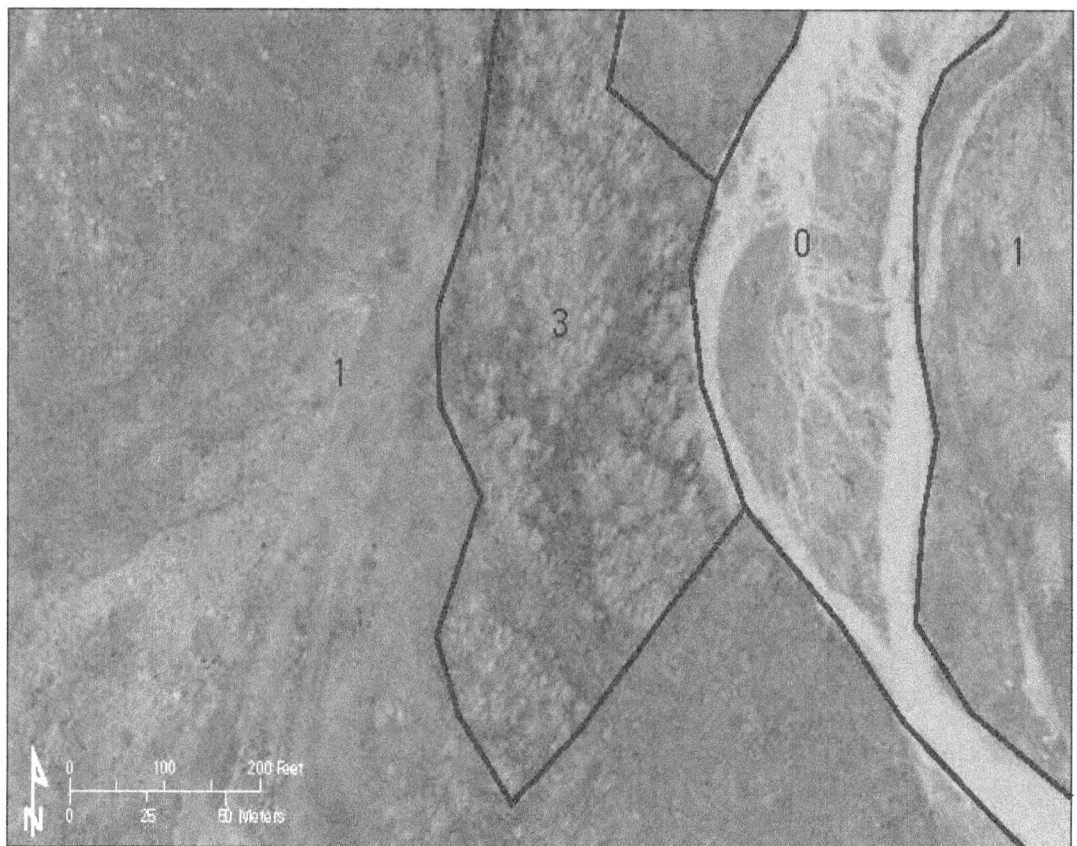

Imagery: USGS orthophotograph, 1-foot pixels, 2006 photography. Projection: UTM zone 6, North American Datum of 1983

Figure 6. Orthophotograph showing typical vegetation for 2006 in the most common age categories near Palmer, Alaska. The left-most category 1 surface is about 30 years old and has ground cover and shrubby vegetation. The category 3 surface contains mostly deciduous trees that appeared similar to the category 1 vegetation in 1949 and is therefore estimated at 87 years old. A slight green tinge along the right side of this surface indicates initiation of spruce where a stand of trees stood in 1949, making this area almost 150 years old. Category 0 indicates active channels and bars too small to define. The right-most category 1 surface is about 16 years old.

Five age categories were established from the three sets of photographs of the full study area (table 5) based on when vegetation first appeared in orthophotographs, and on evidence of channel occupation, usually visible as curved stripes of younger vegetation in former channels and referred to here as channel scars, on surfaces already vegetated in 1949 (fig. 6). Braid plain surfaces that were vegetated in the 1962 photograph, but displayed vegetation typical of 1 to 3 decades of growth in the 2006 photograph, were reclassified to the younger class because it was likely the surface had been disturbed and re-abandoned after the 1962 photograph. Age ranges for these categories (table 5) were assigned from photograph dates and the decadal-scale vegetation succession patterns (fig. 6). The minimum age was assigned as 10 years for vegetation to appear plus the elapsed time since the photograph where vegetation first appeared, and the maximum age was assigned as 9 years of potential vegetation growth plus the elapsed time since the next older, apparently unvegetated photograph. Where vegetation was always present, the maximum age was assigned as the 57 years since 1949 plus the estimated age of the vegetation (40 years for features with channel scars and an arbitrary 143 years for features with no channel scars).

An analysis area was defined as a subset of the braid plain to exclude areas where the braid plain margin was ambiguous. Specifically, braid plain margins for the first several braid plain miles upstream of the river mouth, where the river has shifted course into a low-lying area lacking a defined left bank, were arbitrarily based on evidence of flooding. Analysis of braid plain age and historical erosion omitted this area and used the analysis area shown in figure 7.

Analysis of the braid plain included computation of a simple area-weighted age for 2006 using the area and estimated average age of each feature class. Braid plain bars and islands mapped from the 1949 and 1962 historical orthophotography provided comparative data to look at changes in the composition of braid plain through time. Analysis of these historical products included the relative proportion of vegetated area, computed by summing the areas of the respective braid plain features within each geomorphic reach and dividing by braid plain area.

Braiding Index

The braiding index is a common metric of the intensity of braiding, typically compiled by counting wetted channels across the braid plain and preferred for its insensitivity to channel form and ease of use (Egozi and Ashmore, 2008). For a random point in each reach, 10 transects spaced apart by the active channel belt width, or distance between outermost active channels, were overlain on the orthophotographs for 2006. This metric more closely mirrors the traditional channel width than the braid plain width and exceeded the length scale of most bars. Those same transects were overlain on the

Table 5.   Age range and average age for 2006 braid plain age categories, Matanuska River, southcentral Alaska.

[Average age computed as midpoint of range]

Age category	Description	Age range (years)	Average age (years)
0	Water and unvegetated bars	0–9	4.5
1	Vegetated bar/island not visible in 1962 photograph	10–53	31.5
2	Vegetated bar/island visible in 1962 photograph but not in 1949 photograph	54–66	60
3	Vegetated bar/island with vegetation and channel scars in 1949 photograph	67–97	82
4	Vegetated bar/island with vegetation but no channel scars in 1949 photograph	98–200	149

historical orthophotography for 1962 and 1949, such that the same location was compared from year to year. An additional set of 10 transects was added to reaches 3, 4, and 5, where narrow channel widths resulted in minimal coverage of the reach. Clearwater channels were omitted from the analysis. The braiding index was computed for each reach as the average number of channels conveying mainstem flow.

Historical Bank Erosion

Historical bank erosion was computed as the change in braid plain margin position between selected years of orthophotography. For the purposes of this report, the braid plain margin is referred to as the bank of the Matanuska River. The movement of banks along individual channels within the braid plain was not analyzed for this project. The analysis area for bank erosion (fig. 7) was restricted to areas where the braid plain could be determined. The indeterminate banks in the lower 3 mi of the river, such as those near the Palmer Plant Materials Center, thus were omitted from analysis.

Computation of Erosion between Photograph Dates

The USGS Digital Shoreline Analysis Software (DSAS) version 4.0 (Thieler and others, 2009), designed for analysis of coastal shoreline change, was used for computation of historical braid plain margin change. DSAS is an ArcGIS® extension that automatically generates regularly spaced transects across user-defined banklines and measures erosion or growth at each transect as the distance between successive

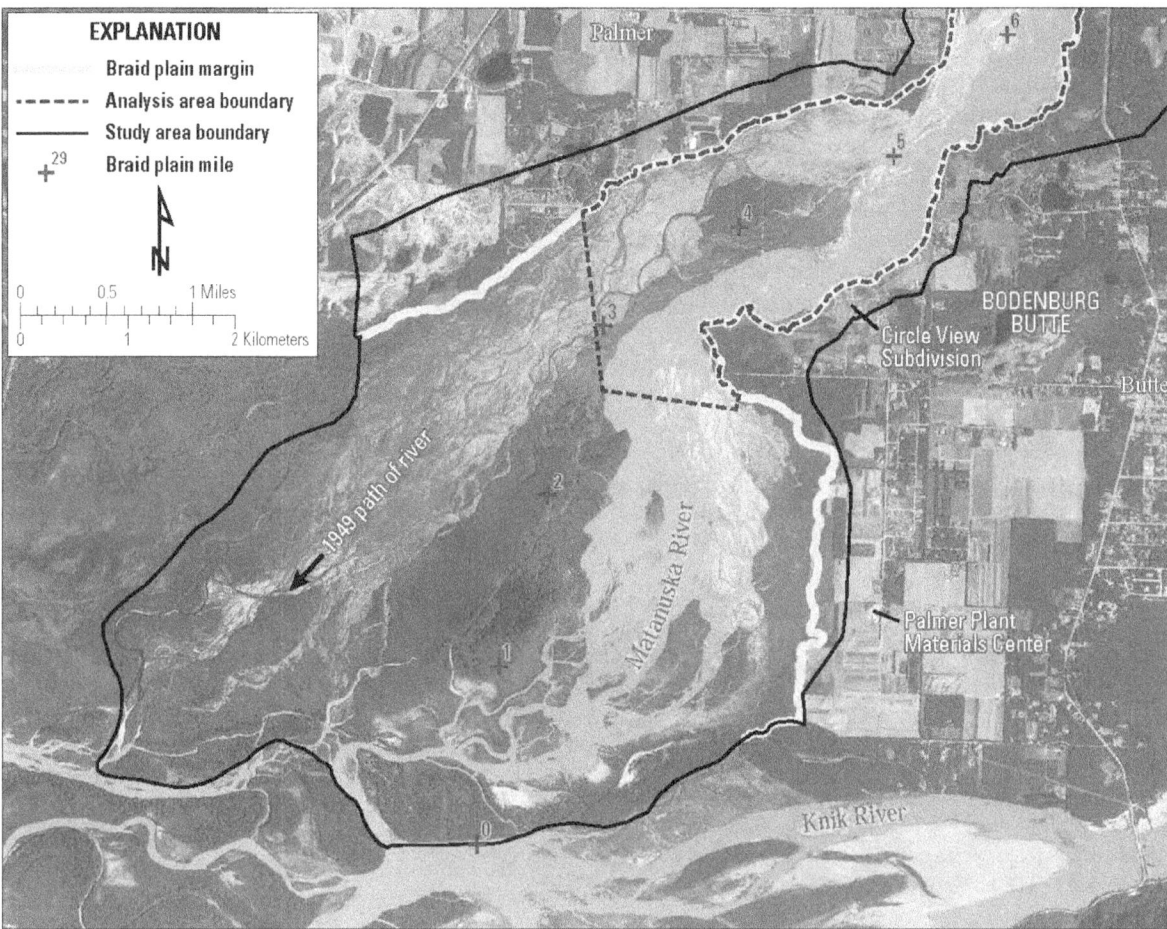

Imagery: USDA-NCRS orthophotograph, 1-meter pixels, 2004 photography. Projection: UTM Zone 6, North American Datum 1983

Figure 7. Detail of boundary differences between the study area, braid plain, and area used for analysis of braid plain age and historical erosion, Matanuska River, southcentral Alaska. The analysis area matches the braid plain area for the rest of the study area.

banklines. DSAS requires a baseline for computation, which was arbitrarily established 50 m (164 ft) beyond the 1949 braid plain margin. Transect spacing of 50 m (164 ft), transect length of 400 m (1,312 ft), and a baseline smoothing distance of 500 m (1,640 ft) provided adequate detail, coverage, and transect angles, respectively. Where banklines curved such that transects intersected banklines twice, transects were shortened or the baseline was adjusted to correct the orientation of the transects. Baselines also were adjusted as needed to allow for braid plain growth at active fans and erosion control feature stabilization.

Computed DSAS statistics include the Net Shoreline Movement (NSM) (change between the youngest and oldest banklines) and the Shoreline Change Envelope (SCE) (total change between all banklines). For river banks with appreciable erosion or growth, the difference is negligible. NSM and SCE were converted to units of feet and a sign convention where positive values represent bank erosion and negative values indicate bank growth. For the purposes of this report, the NSM value representing change between 1949 and 2006 was the most meaningful metric and was used throughout the report to indicate erosion history. For display purposes, 1949–2006 NSM values were converted from transect-based results to a bankline form by segmenting the 2006 braid plain margin in ArcGIS® at the midpoints between transects and transferring each respective transect erosion value to the corresponding line segment.

Historical bank movement (erosion or growth) was computed for the 1949, 1962, and 2006 braid plain margins on the left and right banks for the entire analysis area. A separate analysis of a subsection of the left bank extending from the Old Glenn Highway bridge to the Circle View Subdivision took advantage of additional available orthophotography, increasing resolution to decadal scale intervals.

Error Assessment

The positional accuracy of the geomorphic feature mapping and banklines varied with the registration offsets between the various orthophotography and satellite imagery datasets, determined by comparing control points from 1949, 1962, 2004, and 2006 orthophotography and 2005 satellite imagery at 0.6 mi valley intervals wherever stable features could be identified. Suitable features are sparse and almost no features were consistent throughout all imagery, resulting in variable numbers of control points for any given set of dates (table 6). Control point offset averaged 13 ft and ranged from a minimum of 3 ft for the small area of overlap between the 2004 orthophotography and 2005 Digital Globe satellite imagery to a maximum of 29 ft between the 1949 and 1962 orthophotography (table 6). The offset between the 1949 and 1962 orthophotography appears as a shift, creating the greatest error on banks with a southwest aspect.

Other sources of error in the banklines included the quality and resolution of the orthophotography and errors in interpretation. These compounded errors were estimated by comparing differences between braid plain margins along transects at 50 m (164 ft) intervals on both banks for the length of the erosion analysis area. Values suggesting large magnitudes of braid plain change (erosion or growth) were examined to determine sources of error and were adjusted where correctable errors were present (primarily shadowing). Strong shadowing in the 1949 and 1962 orthophotography obscured features in canyons and along steep banks, particularly on the left bank of the river. Locations of bedrock banks unlikely to have eroded were determined from 2006 orthophotography, which was specifically collected to avoid shadowing, and compared to shadowed banks in other imagery. In a few areas these shadows extend hundreds of feet within the braid plain. In areas where shadowing obviously masked a feature, banklines were adjusted to approximate the actual bank location by extrapolating from adjacent areas and accounting for length of shadows at known features.

Geomorphic features were locally masked by ice and snow cover in selected imagery, which decreased positional accuracy of lines but did not hinder interpretation of features. Other potential errors include poor photograph quality in 1949 and 1962, which resulted in fuzzy signatures that commonly masked the distinction between unvegetated bars within the active channel belt and braid plain bars, and in some cases masked the distinction between braid plain bars and terraces. Interpretation error was minimized by calibrating to field observations and by restricting the line drawing task to one person.

After correcting for shadowing and classification errors, particularly in areas with large braid plain margin changes, the braid plain margin change values were categorized into five classes using the Natural Breaks (Jenks) classification in ArcGIS®. This classification scheme places values into groups based on similarities of the values within the groups, such that banks with significant erosion or growth fell into the outer

Table 6.   Analysis of error in selected orthophotography and satellite imagery datasets using selected control points at 0.6-mile intervals along the Matanuska River, southcentral Alaska.

Dates of imagery compared	Number of control points	Average error between control points (feet)
1949 and 1960	11	29
1949 and 2004	1	22
1949 and 2006	1	13
1960 and 2004	10	12
1960 and 2005	1	7
1960 and 2006	8	9
2004 and 2005	1	3
2004 and 2006	38	10
2005 and 2006	20	9

classes. The class with values between 61 ft of erosion and 83 ft of growth provides a reasonable estimate of the limit of the ability of the data and methods to distinguish actual bank change. The boundaries of this class, which were determined from the dataset, encompass the range of compounded error estimated from the consideration of independent errors. This class contains both error and actual erosion or growth, but because they are indistinguishable in this range, all values were retained. Although 76 percent of the values fell in this class, the net bank change in this class amounted to only 7 percent of the total net bank change in the erosion analysis area, suggesting that erosion summaries were not strongly biased by including these values.

Geomorphology of the Matanuska River

Hydrology and Sediment Supply

Annual maximum instantaneous (peak) streamflows attributed to snowmelt, glacier melt, or rainfall ranged from 12,900 to 46,000 ft^3/s (fig. 8). The dates of the annual peaks fell between mid-May and September, but 74 percent of the peaks occurred in June or July (fig. 3). The maximum annual peak streamflow, 82,100 ft^3/s in August 1971 (fig. 8), resulted from an outburst flood from a moraine- and landslide-dammed lake on a Granite Creek tributary during a regional rainstorm (Lamke, 1972; McGee, 1974). This flood exceeded the magnitude of the published estimate for a non-outburst flood with an exceedance probability of 0.002, equivalent to a recurrence interval of 500 years (Curran and others, 2003). Diel, or daily, fluctuations reflecting glacier melt during daylight periods were routine during the period of record and could be large relative to the total flow. For example, on July 17, 2009, the date of the annual peak, streamflow decreased from the peak of 19,600 ft³/s to a low of 12,600 ft³/s, capping a week when diel fluctuations ranged from 3,000 to 6,000 ft³/s.

Graphs of peak streamflow (fig. 8) and mean annual streamflow (fig. 9) for the period of record do not show clear increasing or decreasing trends, although data gaps limit the strength of this conclusion. The mean peak streamflow (computed as the inverse logarithm of the mean of the logarithms) appeared similar for the subperiods 1949–1962 and 1962–2006. The mean annual streamflow for the subperiod 1949–1962 appeared slightly higher than the mean annual streamflow for the subperiod 1962–2006. However, these comparisons are complicated by the extended periods in the 1970s, 1980s, and 1990s when the streamgage was not operated. In particular, the two values for mean annual streamflow measured during 1974–2000 are low relative to average values. Comparison of mean annual streamflow normalized by the long-term mean for the Matanuska River and for the Knik and Little Susitna Rivers, nearby streams draining solely from the Chugach and Talkeetna Mountains, respectively, shows values below and above the long-term mean annual streamflow for each river (fig. 9). The correlation between rivers is weak (the best match is obtained with the Little Susitna River, r² = 0.29), so the lack of trends in the nearby streams cannot be used to determine if Matanuska River mean annual streamflow during 1974–2000 generally was as low as the two measured values during that period.

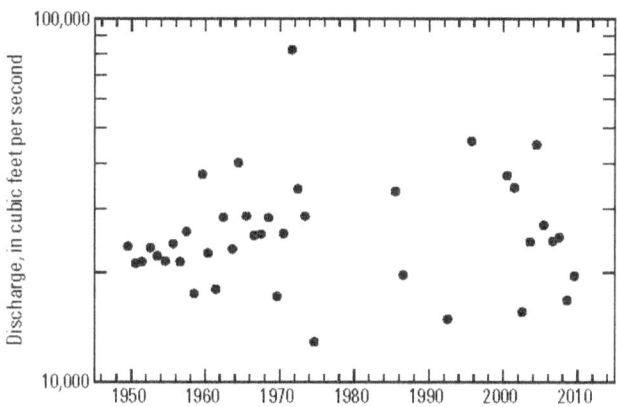

Figure 8.   Annual maximum instantaneous (peak) discharge and year of occurrence during the period of record for the Matanuska River at Palmer (USGS gaging station 15284000), southcentral Alaska. The period of record for peak streamflow includes a peak measured in 1995 when the streamgage was not in continuous operation. Any other large peaks likely would have been documented had they occurred during the discontinued gaging periods (David Meyer, U.S. Geological Survey, oral commun., 2010).

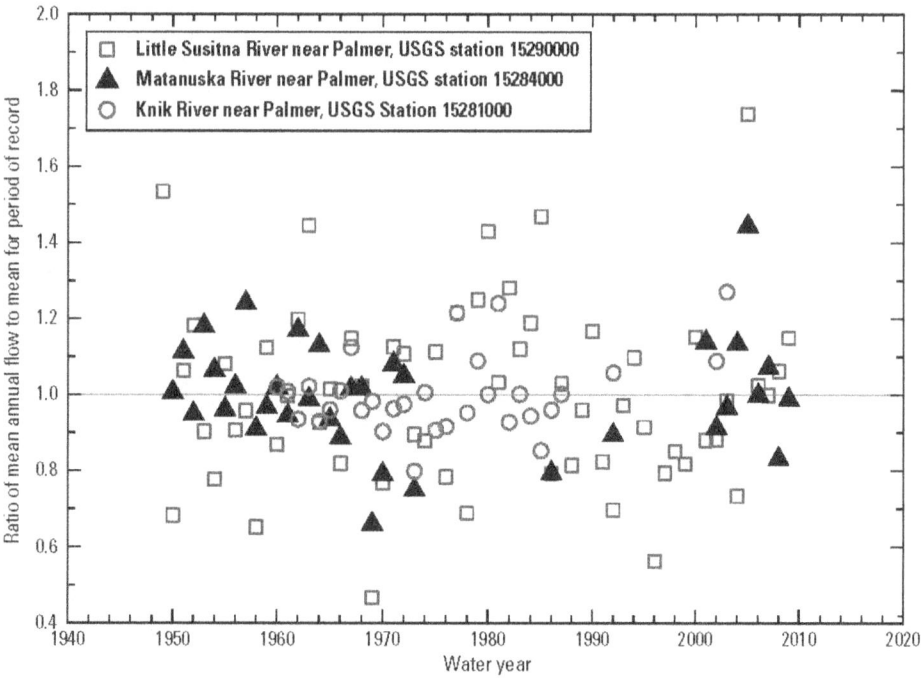

Figure 9.   Ratio of mean annual streamflow to the mean for the period of record for the Matanuska, Knik, and Little Susitna Rivers, southcentral Alaska. The mean for the period of record is 3,880 ft³/s for the Matanuska River, 7,000 ft³/s for the Knik River, and 205 ft³/s for the Little Susitna River.

A rough assessment of the relative influence of tributaries does not show a remarkable dominance of one particular source of water or sediment. A reconnaissance report on the hydrology of Matanuska River tributaries (Restoration Science and Engineering, 2006) measured streamflow once at 15 major tributaries and at South Fork Matanuska River, which can be viewed as the mainstem Matanuska River upstream of the confluence with East Fork Matanuska River. These measurements were not simultaneous but were all collected during a late summer, moderate-to-high flow period. Although only a rough estimate, the results indicate that the sum of tributary inflows were less than one-half of the total Matanuska River flow at the USGS streamgage at Palmer. These results also showed that the mainstem streamflow at Chickaloon was about one-half of the flow at the USGS streamgage at Palmer, 28 mi downstream, on the same day (same result obtained using following day to account for lag time). Concurrent streamflow measurements at the base of the Matanuska Glacier (Denner and others, 1999) and at the USGS streamgage at Palmer show that the maximum instantaneous value for 1995 (the largest non-outburst peak on record) at the glacier was 18 percent of the value at the streamgage.

Future hydrologic conditions could change with climatic conditions, in particular the extent of glacial coverage of the watershed. Within the historical record, the terminus of the Matanuska Glacier has both advanced and retreated. An advance in the 1960s was followed by retreat that continues at the time of this report (Dan Lawson, U.S. Army Corps of Engineers-Cold Regions Research and Engineering Laboratory, oral commun., 2011). Neal and others (2010) computed that the general pattern of glacier thinning and retreat observed across the Knik Arm/Kenai region produced 20 percent of the watershed discharge in the region, suggesting that these elevated discharges are not sustainable with continued melt. Hodgkins (2009) computed that June and July mean monthly streamflow was slightly higher for the Matanuska River (and many other Alaskan glacial rivers) during the warm phase (1977–2006) relative to the cool phase (1947–1976) of the Pacific Decadal Oscillation, a well-documented climate pattern correlated to North Pacific sea surface temperatures. Although specific conditions for the Matanuska Glacier and other glaciers in the watershed were not assessed for this report, these accounts suggest that changes in hydrology could occur, which could in turn contribute to changes in the degree of braiding. Additional

geomorphic changes are possible with potential future outburst floods from existing or future moraine and landslide dams present in the watershed, but the timing and magnitude of those floods is unpredictable.

In addition to streamflow, bed sediment supply is a critical variable in fluvial processes such as braiding, which is generally associated with an ample sediment supply. U.S. Department of Agriculture, Natural Resources Conservation Service (2004) produced a conceptual sediment budget showing bedload production along most of the river, offset by deposition within wide gravel reaches. The Matanuska Glacier produces a large suspended sediment load, but Pearce and others (2003) found that bedload was negligible (less than 1 percent of the total sediment load) at vents of subglacial meltwater and sediment erupting from an overdeepening at the glacier terminus in 2000. They hypothesized from the high percentage of sand and silt frozen to the bed of uplifted parts of the glacier that limited bedload-sized sediment was available for discharge, which would imply that this is not a variable likely to change if streamflow increases. In contrast, they measured substantial bedload transport at Hicks Creek, 7 mi downstream.

While attempting to quantify sediment input was beyond the scope of this report, a qualitative assessment of changes at tributary junctions provided a first-order assessment of the relative importance of tributary inputs to mainstem sediment supply. The wide, braided nature of the Matanuska River upstream of the Matanuska Glacier suggest that other glaciers and streams in the basin headwaters, and the extensive landsliding observed in the East Fork Matanuska River, contribute significant streamflow and sediment. A total of 32 major tributaries entered the Matanuska River in the study area, averaging 2.0 mi apart; an additional 7 major tributaries enter the river upstream. Tributaries were absent in Reach 1, were short and closely spaced in Reach 5, and had larger average watershed areas in the Talkeetna Mountains than in the Chugach Mountains. Extensive, high fans in Reach 5 have been incised by their modern streams, suggesting that substantial fan building occurred during earlier periods when larger volumes of water and sediment were being shed from the Chugach Mountains. Although many of the fans in Reach 5 are truncated, several fans in Reach 3 (see Carpenter Creek in fig. 10C) protrude into the river, providing a potential sediment supply, but also suggesting that the river is not as capable of carrying the sediment provided.

Figure 10.   Geomorphic features along Matanuska River, southcentral Alaska. (*A*) Braid plain and Matanuska River valley inset into broader valley, BPM 28; (*B*) vegetated braid plain, terrace, and Eska Creek fan, BPM 20; (*C*) historically active fan at Carpenter Creek, BPM 31; (*D*) older fan at unnamed tributary, BPM 47.5; (*E*) high uplands (glacial sediments) with bus on Glenn Highway and abandoned railroad near base of bluff, BPM 11; (*F*) clearwater side channel and low bedrock upland, BPM 27.5; (*G*) high bedrock banks forming sharp bend in bedrock canyon, BPM 18.5; and (*H*) bedrock constriction at Old Glenn Highway bridge, BPM 9.

Valley Features

The Matanuska River includes a geologically complex canyon section 2–4 mi wide between the Talkeetna and Chugach Mountains (Reaches 2–7) that transitions at Palmer to a lowland section through unconfined glacial deposits (Reach 1). The complex geology of the Matanuska Valley and the glacial and fluvial reworking of sediment during and following deglaciation were fundamental to establishing the general form of the Matanuska River, a framework unlikely to be modified substantially by the modern river. The outer walls of the valley in places display the classic U-shape of a glacially carved valley, but the floor of the valley is much more topographically complex than a typical glacial valley (for example, the nearby Knik River valley). Short sections of bedrock ridges thinly mantled with sediment protrude parallel to the valley and are interspersed with glacial and fluvial deposits. Interpretation of these deposits is complicated by lack of moraines along the valley and minimal consolidation of glacial sediment suggesting that the glacier did not remain stationary in the valley for long periods (Dan Lawson, U.S. Army Corps of Engineers-Cold Regions Research and Engineering Laboratory, and Sarah Kopczynski, Lehigh University, oral commun., 2008). Recently reported evidence of streamlined features and large fluvial deposits also suggest that megafloods spilled down the valley from paleo-glacial lakes in the Copper River basin (Wiedmer and others, 2010). Large landslides mapped on the flanks of the right valley wall near the Matanuska Glacier (Detterman and others, 1976) likely reached the river at some point, shaping the valley wall and potentially generating outburst floods. For this study, no effort was made to distinguish specific glacial and fluvial deposits not associated with the modern Matanuska River.

The modern Matanuska River is inset into the floor of the Matanuska Valley in a largely bedrock-controlled inner valley less than 1 mi wide. Prominent features of this valley include a bend in the general course of the river at Chickaloon, matching the orientation of faults in the valley, and right-angle bends in bedrock canyons at BPM 18 and 26 (fig. 10). Within the inner valley, here termed the Matanuska River valley, most of the valley floor consists of historical braid plain or modern fluvial deposits formed during or since deglaciation by the Matanuska River (terraces) or by tributaries (fans) (fig. 10). Other features include landslides and glacial deposits. Downstream of Palmer, most of the lowland surrounding the Matanuska River is a glacial outwash terrace. A line of low bedrock knobs, as well as the prominent Bodenburg Butte, help deflect the present-day Matanuska River away from the Knik River valley, which also drains through the lowland. Gravels associated with the Matanuska River found south of

Bodenburg Butte (Fahnestock and Bradley, 1973) and LIDAR data showing fluvially stripped bedrock with a southerly drainage path near Falk Lake, indicate that former courses of the river flowed between at least some of these knobs.

Geomorphic features mapped in the Matanuska River valley floor and some of the side slopes included the braid plain, fluvial deposits, modified land, and selected uplands. All types of mapped features are described in table 4, and the locations of these features and their specific attributes are contained in the GeomorphicFeatures shapefile in appendix A. Figure 11 shows a detail of the mapped geomorphic features.

Features not mapped are considered uplands and shown as a single polygon in the Uplands shapefile (appendix A). Uplands include bedrock, glacial sediments, and other materials not clearly identifiable as fluvial features formed by the modern Matanuska River. Although no geomorphic identifier is assigned to specific parts of the uplands, the Banklines_2006 shapefile (appendix A) provides material and height information.

The study area contains 138 mi of banks, defined for this study as the braid plain margins. Of these, 63 percent consisted of unconsolidated sediment and 32 percent consisted of bedrock (table 7). Bedrock banks are mostly high (84 percent), whereas unconsolidated banks are mostly low (44 percent) but also include significant moderate height (32 percent) and high (24 percent) banks.

Braid Plain

The braid plain consists of areas occupied by the Matanuska River presently or within the historical period (about 100 years). Fundamentally a broad, gravelly plain, it stretches hundreds to thousands of feet across the valley and contains active channels, bars, vegetated islands, and vegetated abandoned channels and bars. For simplicity, short reaches where the channel narrows and the river ceases to braid are included in the following discussions of the braid plain. Braid plain margins for 1949, 1962, and 2006, identified during delineation of banklines and extracted as separate shapefiles, are presented in appendix A.

Vegetation in the braid plain ranged from herbaceous and shrub cover on younger surfaces to mixed spruce and deciduous forest on older surfaces. Large woody debris consisting of isolated entire trees or logs and occasional debris jams are scattered on the braid plain, particularly downstream of eroding forests. Abandoned channels often fill with tributary or groundwater, creating miles of clearwater side channels documented as salmon spawning habitat (Curran and others, 2011).

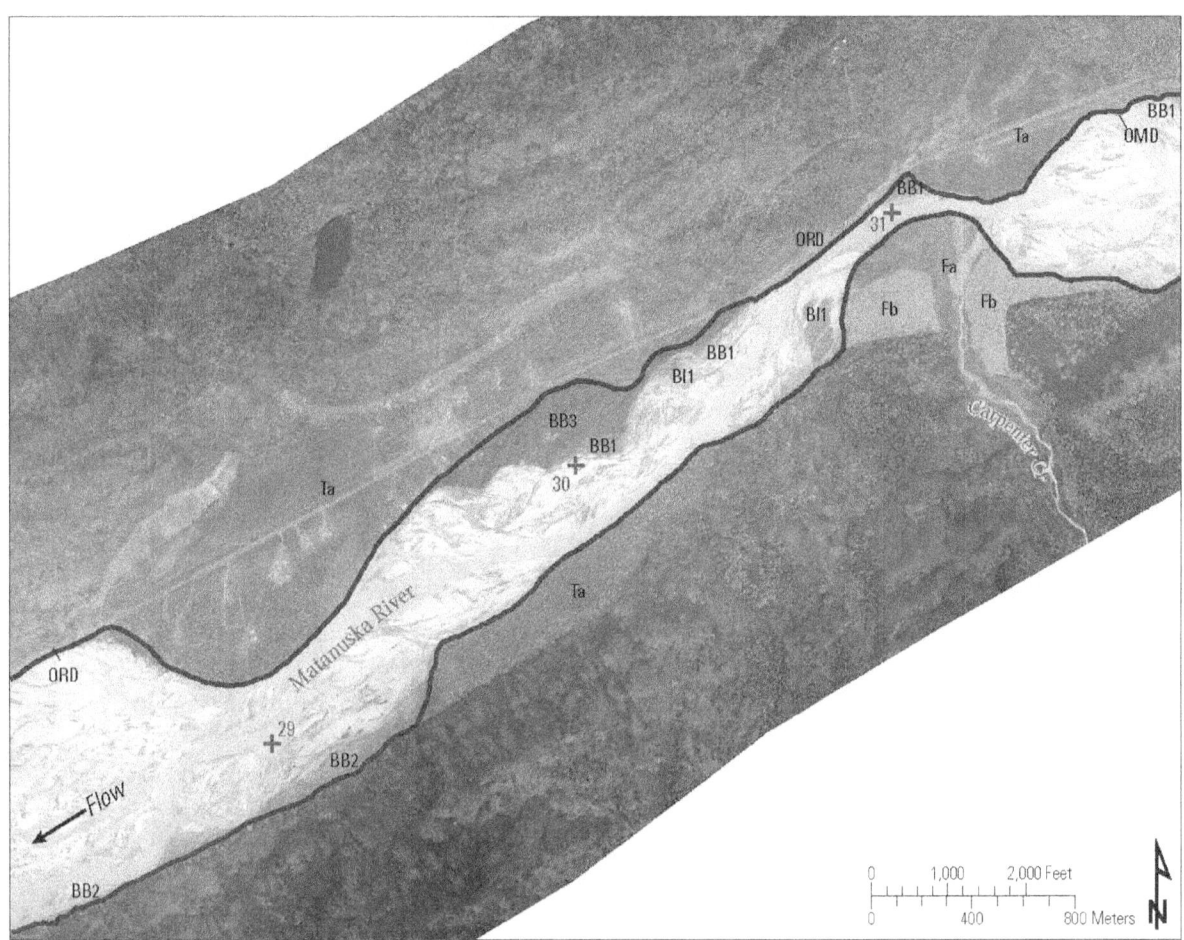

Imagery: USGS orthophotograph, 1-foot pixels, 2006 photography. Projection: UTM Zone 6, North American Datum 1983

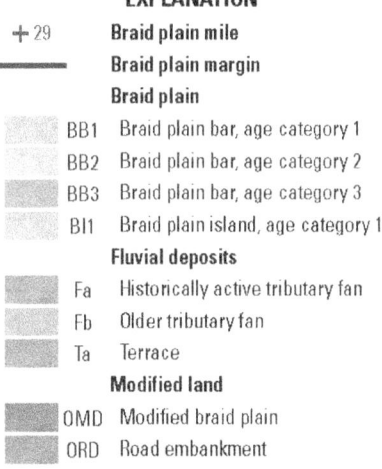

EXPLANATION

+ 29	**Braid plain mile**
——	**Braid plain margin**
	Braid plain
BB1	Braid plain bar, age category 1
BB2	Braid plain bar, age category 2
BB3	Braid plain bar, age category 3
BI1	Braid plain island, age category 1
	Fluvial deposits
Fa	Historically active tributary fan
Fb	Older tributary fan
Ta	Terrace
	Modified land
OMD	Modified braid plain
ORD	Road embankment

Figure 11.   Features along the Matanuska River near Carpenter Creek, southcentral Alaska.

Table 7.   Summary of bank height and material, Matanuska River, southcentral Alaska.

[Banks are defined as braid plain margins. **Abbreviation:** mi, mile]

Bank material	Length of bank				
	Short banks (mi)	Moderate-height banks (mi)	High banks (mi)	Total (mi)	Total (percent)
Artificial fill	3.0	2.3	0.3	5.5	4
Bedrock	1.6	5.4	37.7	44.7	32
Consolidated sediment	0.0	0.0	0.8	0.8	1
Unconsolidated sediment	38.5	27.4	20.8	86.7	63

Braid Plain Width

The width of the Matanuska River braid plain was an average of 2,600 ft in 2006, but varied substantially throughout the study area (fig. 12). Three bedrock narrows and three tributary fans narrowed the braid plain to 130–370 ft wide for short reaches spaced 4–9 mi apart over the downstream one-half of the study area. Intervening areas expanded considerably, creating broad gravel plains 3,000–7,000 ft wide. Reaches 4 and 6 were consistently narrow, averaging 500 and 430 ft wide, respectively (table 8). Braid plain width does not show consistent downstream trends

(fig. 12) that would be expected if changes in streamflow were a strong geomorphic factor. Variability in width is more likely the result of uneven distribution of geological materials within the Matanuska River Valley. Changes in bedrock resistance where Tertiary intrusive volcanic rocks are present near the river probably explain the narrow width of Reach 4. Similarly, coarse morainal materials at the terminus of the Matanuska Glacier severely constrict the channel near Lion Head. Downstream of the braid plain analysis area, the river is not restricted by bedrock and broadens to an expansive plain including former river areas older than the historical period.

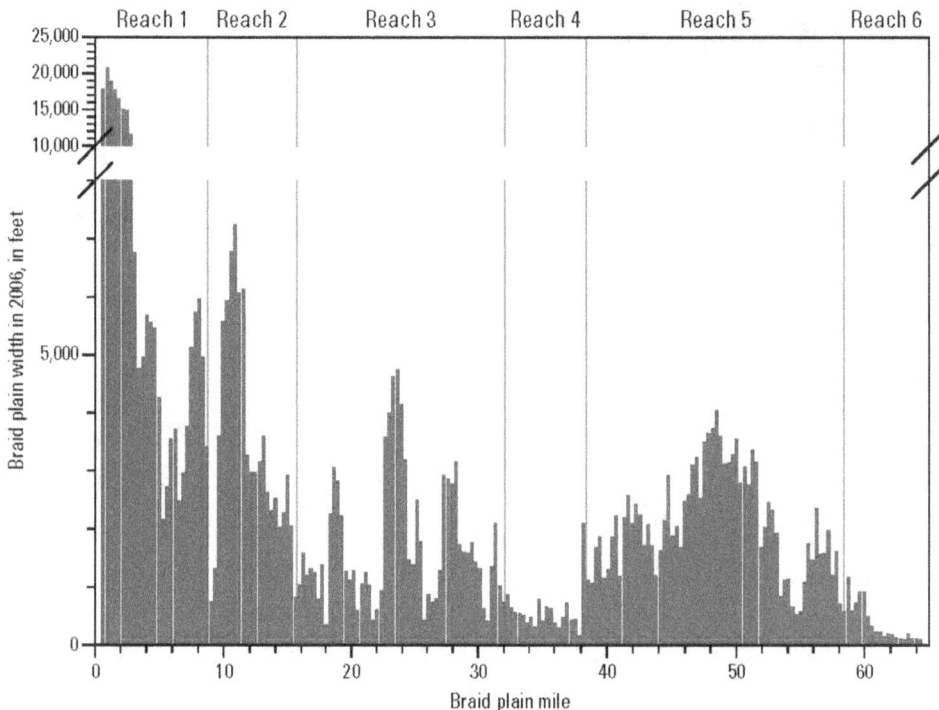

Figure 12.   Braid plain width at 0.3 mi intervals along the Matanuska River, southcentral Alaska.

Table 8. Braid plain and river characteristics in six geomorphic reaches of the Matanuska River, southcentral Alaska.

[Abbreviations: BPM, braid plain mile; mi, mile; mi², square mile; ft, foot; N/A, not available]

| Reach | Descriptive name | Boundaries (BPM) | Braid plain length (mi) | 2006 river length (mi) | 2006 channel planform | Major tributaries | | 2006 braid plain area (mi²) | 2006 average braid plain width (ft) | 2006 river sinuosity | Braiding index | | |
						Number	Number per braid plain mile				1949	1962	2006
1	Lower river	0.0–8.7	8.7	12.1	Braided	0	0	13.7	[1]8,300	1.4	9.1	11.0	8.6
2	Palmer overlook	8.7–15.6	6.9	8.4	Braided	2	0.3	4.3	3,300	1.2	7.2	6.3	6.8
3	Sutton	15.6–32.1	16.5	18.3	Braided[2]	9	0.5	5.3	1,700	1.1	3.6	3.4	4.9
4	Chickaloon	32.1–38.4	6.3	6.5	Single-thread	4	0.6	0.6	500	1.0	1.7	1.0	1.7
5	Gravel Creek	38.4–58.4	20.0	22.1	Braided	16	0.8	7.9	2,100	1.1	3.4	3.3	2.6
6	Matanuska Glacier	58.4–64.9	6.6	6.8	Single-thread	1	0.2	0.5	430	1.0	3.3	2.4	1.1
7[3]	Sheep Mountain	N/A	N/A	N/A	N/A	N/A	N/A	N/A	N/A	N/A	N/A	N/A	N/A
Total in study area			64.9	74.2		32		32.4					
Average							0.5		2,600	1.1	4.7	4.6	4.3
Standard deviation											2.8	3.6	3.0

[1] Braid plain width for Reach 1 includes areas outside the braid plain analysis area.

[2] Braided alternating with short single-thread sections at constrictions.

[3] Outside study area.

Braid Plain Dynamics

Understanding the movement of the active channels of the Matanuska River within the braid plain helps constrain the frequency and persistence of the presence of the channel against the bank, a requirement for the occurrence of erosion. The Matanuska River, except in the single-thread areas of Reach 4 and Reach 6, consists of braided channels that migrate frequently, leaving abandoned channels and bars (braid plain bars) that revegetate until reworked once again. Over time at a single area, or along a short length of the river at a single time, the channel pattern can range from a suite of channels concentrated on one part of the braid plain to isolated channels spanning the entire braid plain, even in areas with braid plain widths up to 4,000 ft.

The high mobility rate of Matanuska River channels has persisted on a multi-decadal scale. The rapidity of the complete reworking of channel patterns near Palmer was documented from decadal-scale aerial photos for 1939–66 by Fahnestock and Bradley (1973), in which they noted capacity for reworking at an annual scale, and reconfirmed by U.S. Department of Agriculture, Natural Resources Conservation Service (2004), where channel patterns compared between April and October 1981 show complete reworking of channel pattern and lateral shifts as much as the full width of the braid plain (Northwest Hydraulics Consultants, 2004, fig. 4). During the present study, a new moderate rapid formed at BPM 39.7 in 2009 when the channel moved laterally and poured through a narrow constriction between small bedrock fins. Channel migration on the order of tens to hundreds of feet from year to year was noted within the braid plain by visual observation and by comparing GPS tracks collected during float trips to the channel patterns in 2006 orthophotography. Channel movement processes observed from these various sources included lateral migration, in which bank erosion creates a shift in the channel thalweg, and avulsion, in which the channel abruptly shifts to a new area of the braid plain without progressively eroding through the banks.

The similarity of most islands in height to the surrounding braid plain suggests that island formation by deposition during particularly large flows (floods) is not a significant process in the Matanuska River. Some islands have concave, scalloped margins, suggesting they are erosional remnants of braid plain bars formed when channels avulsed around them. In other cases, islands may form as vegetation grows on bars in the active channel belt. Some young braid plain bars showed evidence of overtopping by recent flows, suggesting that some vegetation may be stripped without occupation by a main channel, a process that also could lead to isolated patches of older vegetation.

Slopes and channel patterns were examined for indicators of general aggradation or degradation (raising or lowering) of the braid plain. Stable markers are not abundant but include records from the USGS streamgage at Palmer (fig. 2) and the relative position of vegetated surfaces. USGS streamflow records dating to 1950 and rating curves (plots of stage, or water level, versus discharge) since 1991 do not show a trend in stage at low flow that is distinguishable from the interannual variation, which is as much as 8 ft. Relatively high stages recorded during low flows suggest that the channel in the bedrock constriction containing the bridge fills with sediment during low flows and is scoured during high flows. Clear evidence of aggradation in wider braid plain areas was not observed, although repeated measurements would be needed to confirm this. Comparison of LIDAR data for BPM 3 to 6 from 2003 and 2006 (fig. 13) showed the active channel belt widening by 50–600 ft into the braid plain bar along the right bank for about 2 mi (elongated erosional belt in fig. 13) but no conclusive trends of river bed aggradation or degradation. Localized aggradation of 1–6 ft persisted for about 1,500 ft downstream of the spur dikes near Circle View Subdivision.

However, in two locations, the right bank at BPM 4–5 near Palmer and the right bank at BPM 42–43 in Reach 5, an extensive, relatively young braid plain bar surface is situated about 10 ft above the active channel. This provides evidence of local river incision on the order of a total of 5 ft, given typical bank heights within the braid plain. These surfaces can be termed emerging terraces, distinguished from fully established terraces by an upstream connection to the braid plain. Without a historical precedent for this process on the Matanuska River and without understanding the reason for incision, it is premature to assume these emerging terraces will remain stable. The edge of the emerging terrace near Palmer retreated several hundred feet between 2003 and 2006, as described above (fig. 13).

Braid Plain Age

The age of vegetation in the Matanuska River braid plain forms an important historical record of channel mobility. Mapped braid plain features fell into distinct estimated age categories (table 5) on the basis of observations of vegetation succession in orthophotography. Active channels and surfaces with no detectable vegetation (age category 0, up to 9 years old) were most abundant, covering 72 percent of the braid plain (table 9). Of the vegetated surfaces (braid plain bars and islands), the youngest category (age category 1, 10 to 53 years) was most common, covering 23 percent of the braid plain. Only about 5 percent of the braid plain was 54 years or older (categories 2, 3, and 4 combined).

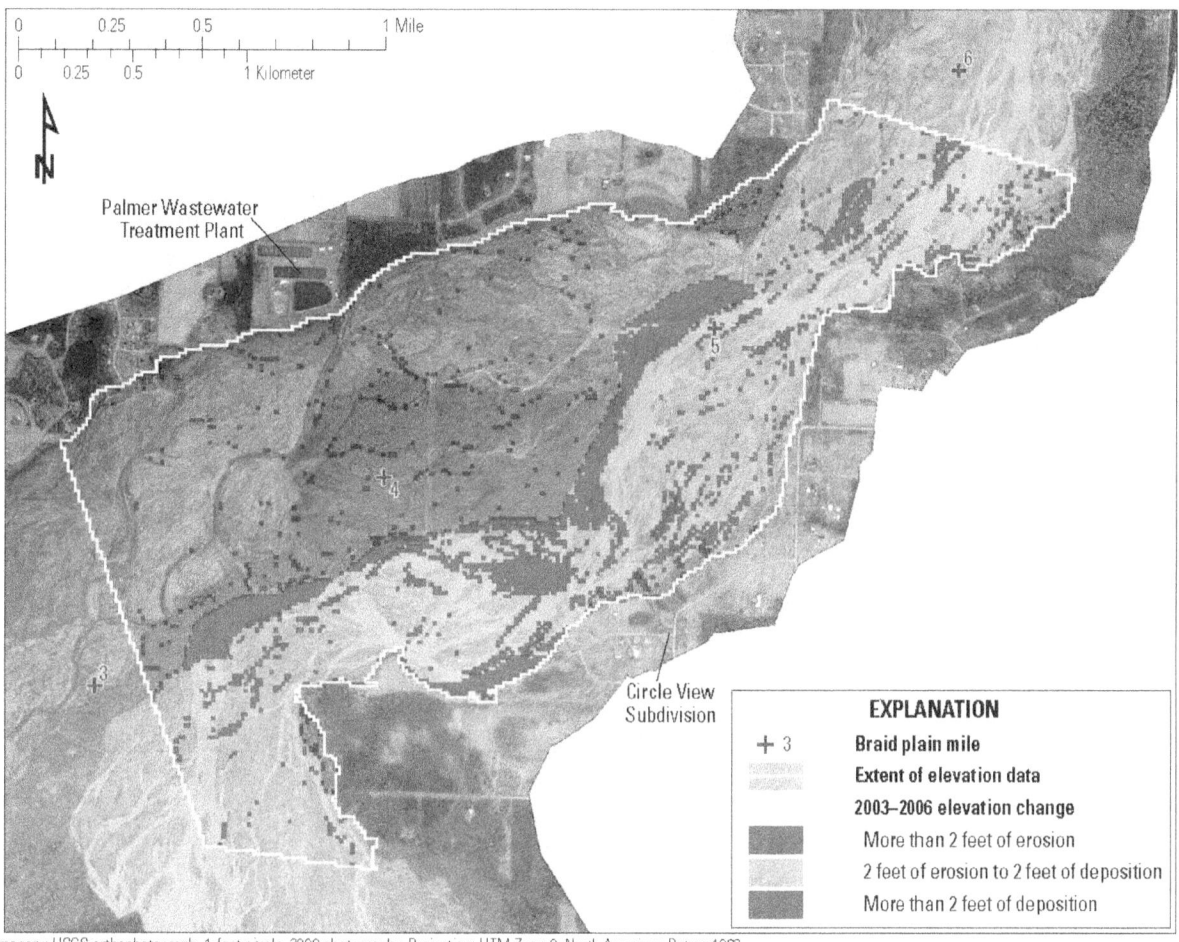

Figure 13.   Change in elevation in the braid plain, Matanuska River, southcentral Alaska, between 2003 and 2006. A slight shift between LIDAR datasets has not been corrected, resulting in spurious paired erosion and deposition values at sharp breaks in slope, such as along the spur dikes near Circle View Subdivision.

Distribution of vegetation also highlights the active nature of the Matanuska River. Water and unvegetated bars comprised 72 percent of the 2006 Matanuska River braid plain, vegetated braid plain bars covered about 25 percent, and vegetated islands occupied only about 3 percent (table 9). The rare occurrence of islands implies that a suite of active channels migrated frequently across the braid plain and that vegetation did not appreciably limit channel movement.

Vegetation appears to have increased within the historical record. The percentage of the entire braid plain vegetated was constant at 13 percent between 1949 and 1962, but doubled between 1962 and 2006 (table 10). Some of the apparent increase might be the result of the improved resolution, limited shadows, and true color of the 2006 orthophotography, but inspection by reach shows that the changes are not evenly distributed as would be expected if image quality was an overriding factor. Vegetation cover decreased slightly in Reach 3, became 5 times greater in Reach 6, and increased an amount similar to the average in other reaches (table 11). Reaches 1 and 5 are the largest reaches and together accounted for 77 percent of the vegetation along the river in 2006, but Reach 1 accounted for a disproportionately high percentage relative to its area (54 percent of the braid plain vegetation in 42 percent of the braid plain area).

The area-weighted age of the 2006 Matanuska River braid plain was computed as 15 years, using average ages estimated from vegetation visible in orthophotography. A sensitivity analysis using the possible range of vegetation ages bounds the area-weighted braid plain age at 5–24 years (table 12). These results indicate that over the past 150 years, surfaces in the braid plain have been stable on average for only a few decades before being reworked by the river. Although this cannot be used to predict future channel occupation at a particular location, it is a critical measure of general expectations of frequency of channel occupation.

Table 10. Braid plain area and amount of vegetation cover in the Matanuska River braid plain, southcentral Alaska, 1949, 1962, and 2006.

[**Abbreviation:** mi^2, square mile]

Year	Braid plain area (mi^2)	Area vegetated (mi^2)	Percentage of braid plain vegetated
1949	22.2	3.0	13
1962	22.7	2.9	13
2006	23.4	6.6	28

Table 9. Area of braid plain by feature type and age category during 2006, Matanuska River, southcentral Alaska.

[N/A, not applicable, <, less than]

Age category	Area of 2006 braid plain	
	(acres)	(percent of total)
Water and unvegetated bars		
0	10,761	72
Braid plain bars		
1	3,002	20
2	90	1
3	545	4
4	36	<1
Braid plain islands		
1	523	3
2	2	<1
3	28	<1
Modified land		
N/A	15	<1

Table 11. Amount of vegetation cover of the Matanuska River braid plain, southcentral Alaska, 1949, 1962, and 2006, by geomorphic reach.

Year	Percentage of reach vegetated					
	Reach 1	Reach 2	Reach 3	Reach 4	Reach 5	Reach 6
1949	14	18	16	9	9	12
1962	15	18	13	11	9	11
2006	40	27	14	25	29	59

Year	Percentage of vegetated study area braid plain contained in reach					
	Reach 1	Reach 2	Reach 3	Reach 4	Reach 5	Reach 6
1949	45	17	19	1	16	1
1962	47	18	16	1	17	1
2006	54	11	7	1	23	3

Table 12. Area-weighted age of the 2006 Matanuska River braid plain, southcentral Alaska.

Feature	Area (acres)	Area-weighted age	
		Range	Average
Water and unvegetated bars	10,761	0–9	4.5
Braid plain bars	3,673	20–61	41
Braid plain islands	553	13–55	34
Braid plain total	**14,987**	**5–24**	**15**

Channel Characteristics

Matanuska River bed material generally can be described as sand and gravel with varying amounts of cobbles (fig. 14). Boulders were present locally, and silt drapes were common over abandoned channels. A slight general downstream fining was observed over the study area. Median particle size (D_{50}) at the heads of two active bars (recently submerged bars near the water edge) in Reach 5 was 2.4 in. (at BPM 41.3) and 2.2 in. (at BPM 49.0, fig. 14A), and classified as very coarse gravel. The U.S. Department of Agriculture, Natural Resources Conservation Service (2004) collected particle-size measurements along two cross sections in Reach 1 and described the bed material as predominantly gravel (D_{50} of 0.5 in.) with cobbles only along the active channel. Field observations documented a significant coarsening of bed material in Reach 4, including bars containing subrounded, imbricated boulders (fig. 14B) suggesting deposition at large flows. Hydraulic forces and the relative influence of floods from large tributaries like the Chickaloon River are likely greater in this narrow, single-thread channel reach, which may account for the coarser bed, but it also may be a function of the more resistant bedrock in this reach. Boulders present in Reach 6 are outsized relative to bed sediment and are likely morainal material. An isolated boulder deposit on the left bank at BPM 54, upstream of Gravel Creek, had no sand or gravel exposed at the surface; contained imbricated, subangular cobble-to-boulder sized clasts; and stood 5–10 ft above the adjacent braid plain, suggesting deposition by a local, outsized flood.

Figure 14.   Bar material along the Matanuska River, southcentral Alaska. (A) Reach 5, braid plain mile (BPM) 49.0, largest opening is 180 mm (7.1 in.). (B) Reach 4, BPM 36.3, notebook is 8 in. long. (C) Reach 3, BPM 24.0. (D) Reach 1, BPM 4.2.

Bank material in unconsolidated sediments was mostly sand and gravel, with cobbles common but decreasing in amount downstream. Boulders were present in selected locations, such as at coarse fans like Granite Creek. Sandy banks were observed in selected locations between Kings River and Granite Creek and extensively on the left bank downstream of BPM 2. No fine-grained deposits typical of backwater conditions were observed at major tributary confluences along the Matanuska River banks exposed at river level.

The slope of the Matanuska River, as determined from topographic map contours and the 2006 river centerline, averages 0.004 over the study area, increasing gradually from 0.003 near the river mouth to 0.006 near the Matanuska Glacier (fig. 15). This is several times steeper than the adjacent Knik River (Fahnestock and Bradley, 1973). At the scale of available topographic mapping, no abrupt changes in slope were detectable at the transition from confined canyon to unconfined lowland near Palmer, the single-thread reach near Chickaloon, major tributaries, or any other probable location. The lack of pronounced steepening at the head of the profile in figure 15 is because the study area ends tens of miles below the mountainous river headwaters.

The sinuosity of the 2006 Matanuska River centerline, drawn along the largest individual channel, peaks at 1.4 in Reach 1 and decreases upstream (table 8). Sinuosity is 1.1 or 1.0 for the upper 76 percent of the river, which indicates that individual braids in this area do not exhibit a strong meandering character. The computed 2006 sinuosity values represent conditions at low streamflow and could decrease at higher streamflow.

The braiding index for the river between the Matanuska Glacier and the river mouth averaged 4.3 in 2006 and appears to have declined slightly since 1949 (table 8). Within individual reaches, the braiding index ranged from 1.0 to 11.0. The only notable historical trend was in Reach 6, where the braiding index decreased from 3.3 to 1.1 during 1949–2006. This decrease is corroborated with observations from orthophotographs of a narrowing of the active channel belt and a corresponding shift from a braided to a single-thread channel near the Matanuska Glacier (fig. 16). This change is localized to the reach including the glacier and was observed in both the Matanuska River mainstem downstream of the glacier and Glacier Creek downstream of the glacier, suggesting a possible correlation to changes in glacier sediment or water discharge. If this is true, the lack of strong channel pattern changes elsewhere suggest that the hydrologic influence of changes in the glacier on this scale is local.

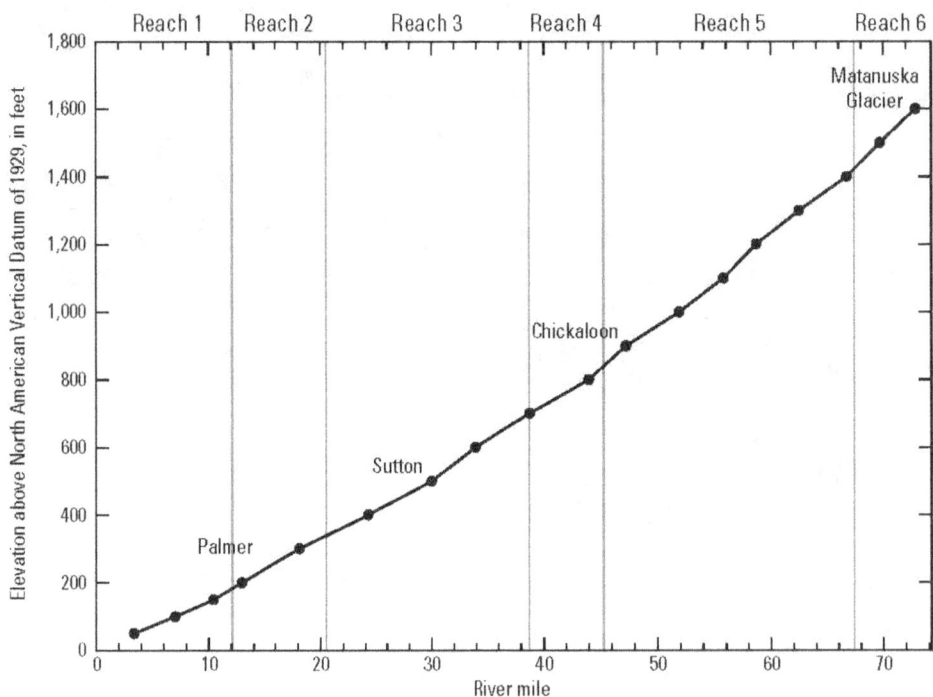

Figure 15.   Longitudinal profile of the Matanuska River, southcentral Alaska, from USGS topographic map contours and the 2006 river centerline.

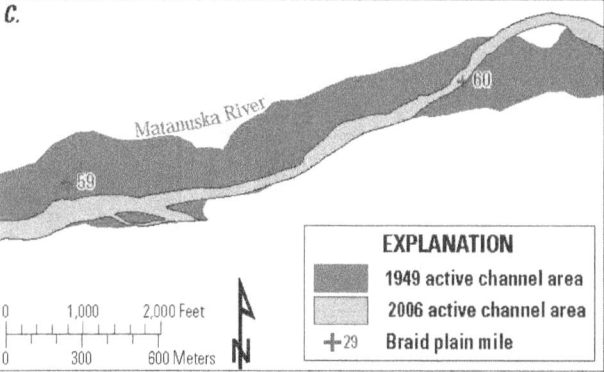

EXPLANATION

1949 active channel area
2006 active channel area
+29 Braid plain mile

0   1,000   2,000 Feet

0   300   600 Meters

Basemap modified from USGS digital datasets, various scales. Imagery: USGS orthophotograph, 1-foot pixels, 2006 photography and USGS orthophotograph, 2-foot pixels, 1949 photography. Projection: UTM Zone 6, North American Datum 1983

Figure 16. Channel pattern changes near the Matanuska Glacier terminus, southcentral Alaska. Progression from a braided channel in 1949 (*A*) to a single-thread channel in 2006 (*B*) and narrowing of the active channel belt (*C*). Only Matanuska River area is mapped, but changes are similar in Glacier Creek.

Bank Erosion

Erosional Processes

Bank erosion is defined here as expansion of the braid plain, and typically occurs along the Matanuska River as fluvial erosion to the toe of a bank, followed by collapse of overlying material into the river and subsequent transport of the material downstream and away from the bank. This lateral erosion is in contrast to flooding, a separate hazard in which water spills over the tops of the banks and inundates land outside the braid plain but returns to the braid plain when water levels recede. Areas where this was observed in the historical period were designated historically flooded areas.

Unconsolidated sediments found along the Matanuska River were almost entirely noncohesive, lacking intergranular strength that would hold blocks of material together and form persistently undercut banks. Observations during a period of active erosion near Circle View Subdivision in 2006 showed that large foot-deep sections of bank failed spontaneously from undercutting, leaving a vertical face. A deep channel adjacent to the bank efficiently transported this material, leaving no accumulation above the water surface. Slumping and rotational failure was not a common process along the Matanuska River. Springs or other evidence of excessive soil moisture, suggesting bank failure by mechanisms other than grain-by-grain removal and ravel or collapse of overlying material, were observed in a few isolated instances, the most obvious at BPM 60 where a spring emerged from active landslide deposits observed within the boundary of landslide mapped by others.

Bank erosion generally converts banks irretrievably to braid plain. The elevation of the land is lowered and the river shifts to occupy the former bank. When the river moves away again, an abandoned channel or bar is left where the former bank stood. Even though this abandoned surface revegetates, it is part of the braid plain and can be reoccupied swiftly by the river. The braid plain can only narrow again by deposition at tributary fans or by creation of terraces. If the river incises and remains at a new, lower elevation for an extended period of time, adjacent braid plain can be considered a new terrace. No permanent terraces appear to have been created in the historical period, but two young braid plain surfaces, on the right bank at BPM 4–5 near Palmer and BPM 42–43 in Reach 5, were observed to be higher than the present river level and could eventually become terraces if the river elevation remains low and the channel does not erode into them.

Changes to a 1-mi reach between Kings River and Granite Creek illustrate the basis for using the historical braid plain as the baseline for measurement of erosion (fig. 17). An abandoned suite of channels formed a surface along the right bank that appears lightly vegetated in 1949 and was stable until 1962, when it appeared more heavily vegetated. By 2004, the active channels migrated across the braid plain through most of the now forested surface and began eroding the braid plain margin in a few places. By 2010, only a few small remnants of the forested surface remained. Measurements from orthophotographs and field observations show channel migration through forested braid plain of up to 100 ft toward the right bank between 2004 and 2006 and up to 500 ft between 2006 and 2010. Erosion to the low terrace at the braid plain margin totaled about 50 ft in selected locations by 2010, damaging residential properties and the Glenn Highway. The clear indicators of recent channel occupation along the right bank in the 1949 image guided the definition of the 1949 braid plain margin The swift reoccupation of the braid plain by 2006, but slower erosion of the braid plain margin, exemplifies the braiding and erosion processes observed on the Matanuska River.

Historical Bank Erosion

Net braid plain margin change from 1949 to 2006, presented as transect-based values projected onto banklines in the Erosion shapefile (appendix A), was used as the primary basis for analysis of historical bank erosion. Class boundaries determined from the data were applied to reduce the dataset to convenient categories. An initial five-class Natural Breaks (Jenks) classification in ArcGIS® used during error assessment provided boundaries for a growth class; an indeterminate class containing change that did not exceed the cumulative user and photography related errors; and three erosion classes. The two upper erosion classes were combined into a single class on the basis of a two-class Natural Breaks (Jenks) classification of the entire dataset that suggested a boundary of severe erosion approximately equal to the lower boundary of the middle erosion class. The resulting final classes were labeled growth, indeterminate change, minor erosion, and severe erosion. The term erosion hotspot is used interchangeably with area of severe erosion for this report. Figure 18 shows the class boundaries and an example of the categorized banklines for an area along the Old Glenn Highway near Palmer.

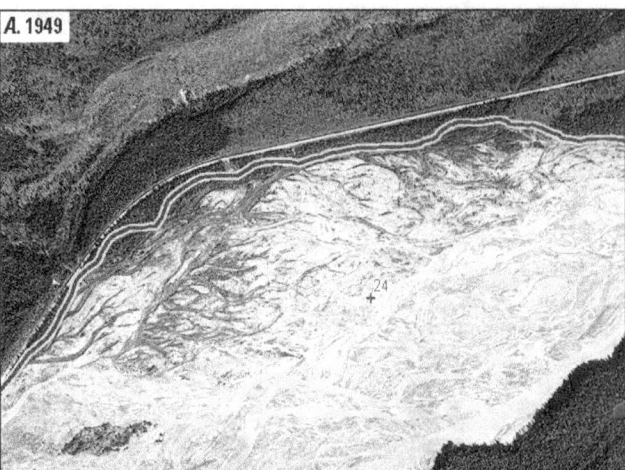

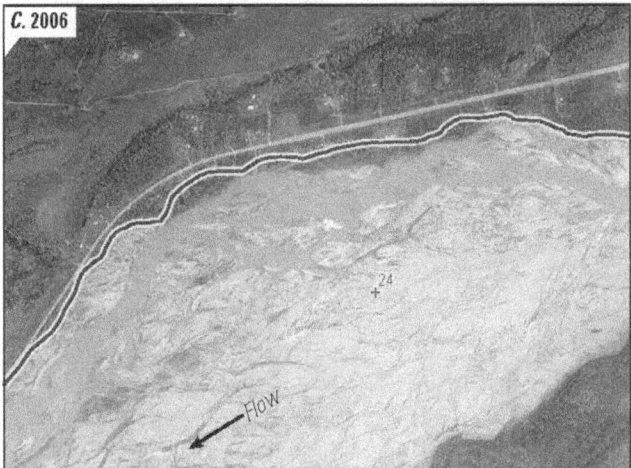

Basemap modified from USGS digital datasets, various scales. Projection: UTM Zone 6, North American Datum 1983

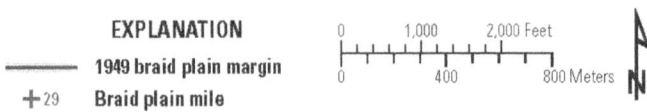

Figure 17.   Braid plain change 1949–2006 between Kings River and Granite Creek, southcentral Alaska. The 1949 braid plain margin line is shown in each image.

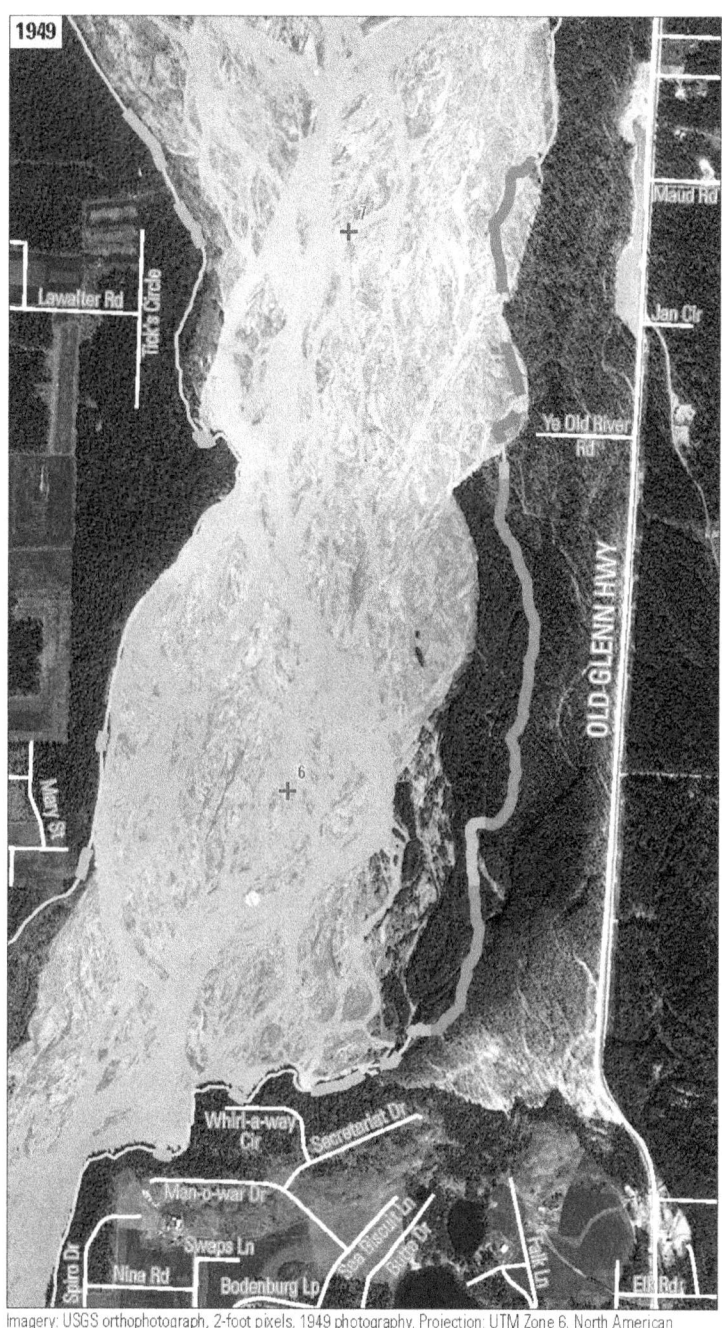

Imagery: USGS orthophotograph, 2-foot pixels, 1949 photography. Projection: UTM Zone 6, North American Datum 1983. Basemap modified from USGS digital datasets, various scales.

Figure 18.   Braid plain margin change classified as growth, indeterminate change, minor erosion, and severe erosion, Matanuska River, southcentral Alaska, 1949–2006.

Bank erosion along the 133 mi of braid plain margins within the erosion analysis area converted a total of 861 acres to braid plain between 1949 and 2006, but the distribution of erosion was localized. Erosion detectable with the project methods occurred at an average rate of 0.9 ft/yr (fig. 19A) from 1949 to 2006 but affected only 23 percent of the banks. Severe erosion accounted for 64 percent of all erosion but occurred at only 8 percent of the banks.

Compiling erosion by reach helped quantify the variable distribution of erosion (figs. 19 and 20). Reaches 4 and 6 were relatively stable, affected by only minor erosion on 6 percent or fewer banks. Severe erosion was most concentrated in Reach 1, affecting 20 percent of the banks. Although a small amount of severe erosion occurred in Reach 3, minor erosion was as prevalent there as in the reaches with more severe erosion.

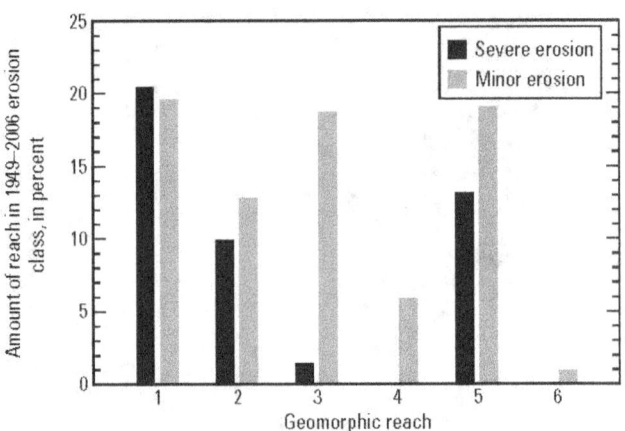

Figure 20.   Minor and severe bank erosion by geomorphic reach, Matanuska River, southcentral Alaska, 1949–2006.

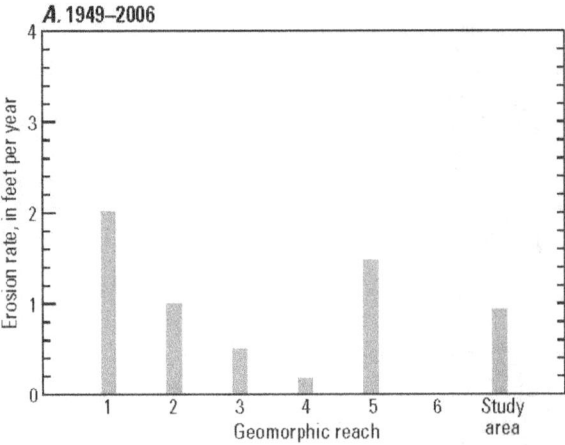

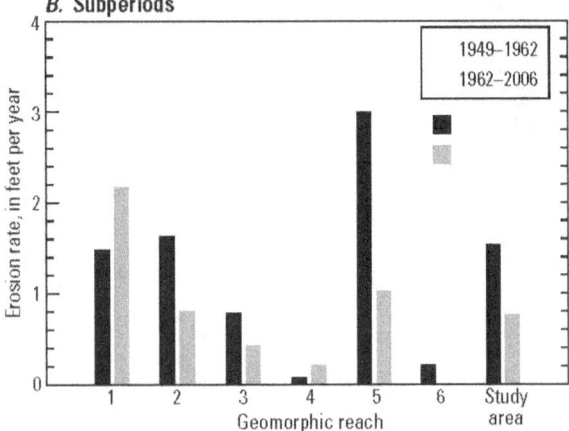

Figure 19.   Bank erosion rates by geomorphic reach, Matanuska River, southcentral Alaska, for (A) the entire study period 1949–2006 and (B) the subperiods 1949–1962 and 1962–2006.

Severe erosion was clustered into 20 erosion hotspots totaling 10.8 mi of bank (fig. 21). Two additional areas were present but omitted from the following analysis because shadowing on orthophotography limited confidence in their interpretation. Left bank erosion hotspots were longer (an average of 4,300 ft compared to 1,600 ft) and more severe than right bank erosion hotspots. Manual identification of eroded features, which varied from mapped feature identification if the feature was eroded away, showed hotspots most often impacted modern fluvial terraces (45 percent of hotspots) and tributary fans (40 percent), followed by glacial terraces (15 percent). This indicates that the river is actively eroding older features (tributary fans and glacial terraces) as well as reworking modern fluvial deposits.

Growth of the braid plain occurred where tributary streams debouched tributary fan material onto the braid plain (denoted in geomorphic features mapping in appendix A as braid plain fans). This included overflow at Granite Creek from the 1971 flood, and fan building events at small tributary streams at BPM 26.6 and 26.9 upstream of Kings River that appear to have been within the last decade but were not investigated for this study. The growth category also includes small reclaimed areas where erosion control was placed in the braid plain, such as at Maud Road/Ye Old River Road (fig. 18). Strong shadowing in Reaches 5 and 6, where older vegetated braid plain bars adjacent to the features make it unlikely that the banks eroded, resulted in a few banks misclassified as growth of up to about 200 ft.

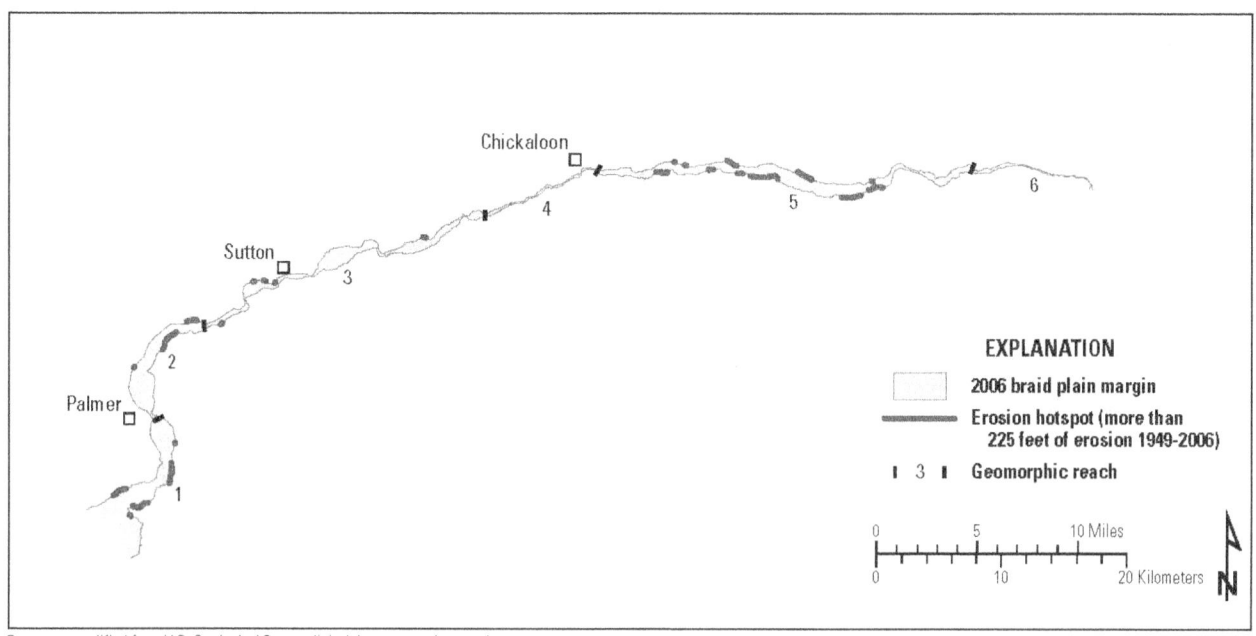

Base map modified from U.S. Geological Survey digital datasets, various scales.
Projection: Universal Transverse Mercator Zone 6N, North American Datum 1983

Figure 21. Erosional hotspots along the Matanuska River, southcentral Alaska, 1949–2006.

Erosion and Bank Characteristics

A summary of net area eroded between 1949 and 2006 by material category shows that as expected, most erosion (85 percent of the total 861 acres) occurred at unconsolidated sediment banks. Values attributed to bedrock banks totaled 13 percent of the area but inspection of these areas indicates much of this is mild erosion along the left bank, where shadows affected the accuracy of bankline delineation. The apparent erosion assigned to bedrock at these areas could be erroneous or could be the result of small braid plain bars completely eroded away. For the purposes of assessing future hazards, erosion to bedrock, artificial fill, consolidated sediment are expected to be minimal based on the resistance of the material, so the remaining analysis focuses on unconsolidated sediment.

A summary of erosion statistics by height category for unconsolidated sediments shows that the average erosion value at a transect on low-height banks was similar to erosion at moderate-height banks (fig. 22A), but about one-half of the area eroded occurred at low-height banks (fig. 22B). This confirms that the abundance of low-height banks and their higher erosion rates places them in the highest risk category for the entire river corridor, but suggests that moderate-height banks offer only modest additional erosion resistance when erosion at a particular location is considered. Similarly, several high banks in Reach 1 had confirmed measurable erosion, demonstrating the lack of significant erosion resistance of unconsolidated banks in the 20–30 ft height range.

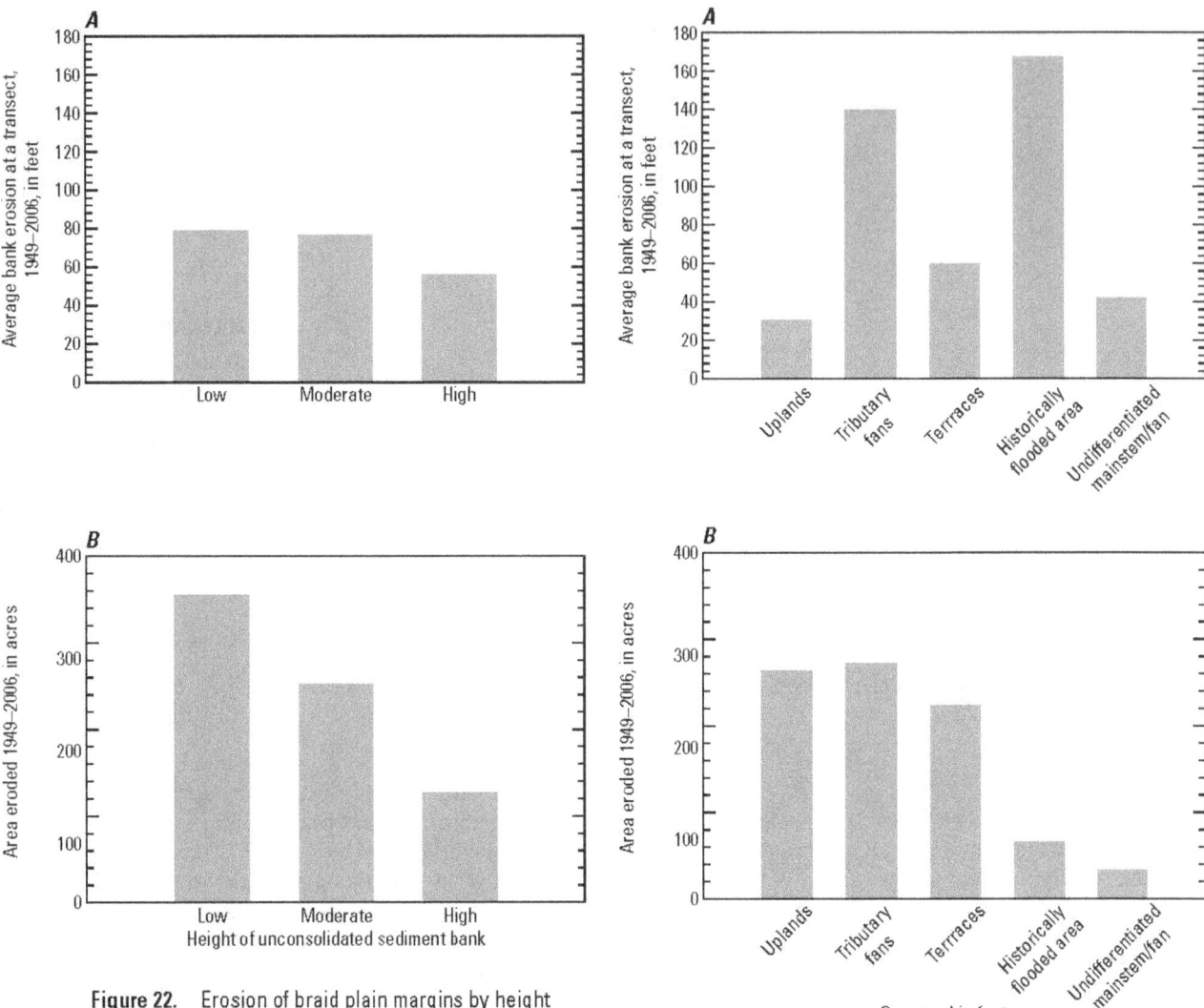

Figure 22.   Erosion of braid plain margins by height category, Matanuska River, southcentral Alaska, 1949–2006. (*A*) Average braid plain margin change (bank erosion) at a transect, and (*B*) net area eroded.

Figure 23.   Erosion of braid plain margins by geomorphic feature, Matanuska River, southcentral Alaska, 1949–2006. (*A*) Average bank erosion at a transect, and (*B*) net area eroded.

Erosion and Geomorphic Features

The distribution of all magnitudes of erosion by geomorphic feature category indicated that the average erosion at a transect was 1.4 to 6.5 times greater (fig. 23*A*) and the total area of erosion was 2.3 times as great along fluvial deposits (primarily tributary fans and terraces) as along uplands (older features including glacial terraces) (fig. 23*B*).

Fluvial deposits of all ages accounted for 43 percent of the bank length but 69 percent of the erosion between 1949 and 2006. Of the fluvial deposits, fans eroded 2.3 times the rate of terraces (fig. 23*A*) but were less abundant, resulting in a similar total amount of erosion (fig. 23*B*). Eroded uplands were more common in the lower half of the river, particularly in Reach 1 where glacial terraces were extensively eroded.

Erosion Persistence

Two subperiods from the three sets of primary analysis orthophotographs provided an opportunity to quantify variability in erosion rates, and a third subperiod using 2004/2005 and 2006 orthophotographs provided a qualitative assessment of short-term change at erosion hotspots. Although these subperiods are not ideal because of their variable lengths, they provide a sense of spatial persistence of erosion and correlation of erosion to hydrologic conditions. River-long erosion rates were greater for 1949–62 than for 1962–2006 (fig. 19B) in every reach except Reach 1, despite the exceptionally large flood in 1971. Of the 20 erosion hotspots, only 9 were eroded in both 1949–62 and 1962–2006 subperiods, and only the hotspot near Circle View Subdivision in Reach 1 had detectable erosion in the subperiod 2004–06. Historical survey maps (U.S. Surveyor General's Office, 1915 and 1916) provide historical context for the pronounced erosion at Circle View Subdivision (fig. 24), where erosion has been persistent and severe since at least 1916. The variability of timing of historical erosion indicates a poor correlation between erosion and flooding.

Additional orthophotography available in Reach 1 permitted a specific analysis of erosion persistence. A linear regression of braid plain margins positions for 1949, 1962, 1985, 1990, 2004, and 2006 was computed in DSAS for the left bank from BPM 3 to 9. Persistence can be represented by a high coefficient of determination (r^2), which measures the fit of the regression line to the data. If erosion rates are nearly constant, the regression will closely match the data. For most transects in the two major erosion hotspots in Reach 1, the areas near Circle View Subdivision and along the Old Glenn Highway south of the Maud Road/Ye Old River Road erosion control, r^2 fell into the 0.82 to 0.97 category, confirming that erosion here has been persistent since 1949. No other areas within this reach had r^2 values in this highest category.

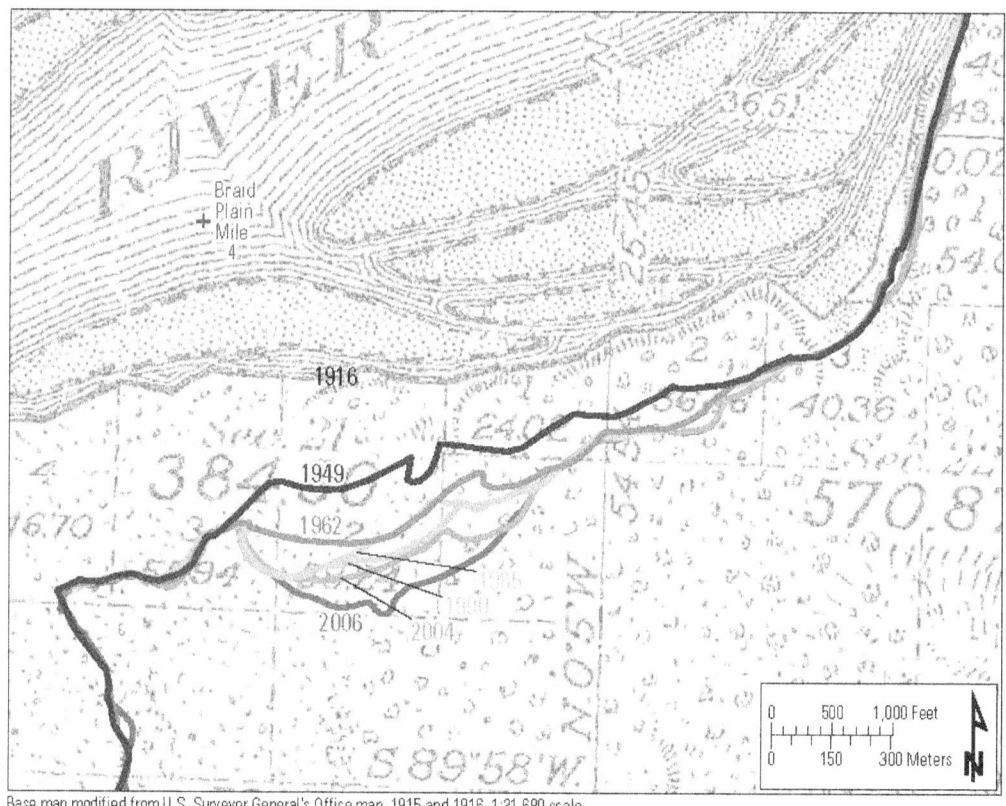

Base map modified from U.S. Surveyor General's Office map, 1915 and 1916, 1:31,680 scale

Figure 24. Braid plain margin position between 1916 and 2006 near Circle View Subdivision, Braid Plain Mile 4 of the Matanuska River, southcentral Alaska.

Shift in River Course near River Mouth

Prior to 1962, the lower 3 mi of the Matanuska River consisted of a suite of braided channels along the right bank and a broad plain along the left bank, near the Palmer Plant Material Center (fig. 7). This broad plain was not substantially elevated above the river but was fully forested without distinct evidence of recent mainstem occupation. For this study, this area can be best described as flood plain because it may have supported some overtopping by the mainstem without channel occupation, and because it lacks the distinct boundaries of the historical braid plain elsewhere. This area likely was occupied by the mainstem of the Matanuska River hundreds to thousands of years ago, and parts of the area were inundated by Knik River glacial outburst floods prior to 1963.

A pronounced shift moved the river from a southwest course discharging to Cook Inlet distributaries in 1917 and 1949 to a southerly course discharging predominantly to the Knik River by 1996 (see indication of former channel pattern in fig. 7). The left bank from BPM 3 to 5 eroded hundreds of feet between 1917 and 1949, and was stopped only by an isolated bedrock point at BPM 3.5. By 1962, mainstem water invaded a pre-existing network of small channels in the low-lying flood plain area downstream of this bedrock point. By 1985, a distinct and enlarged pre-existing channel was conveying mainstem flow. By 1990, the upper one-half of this channel had expanded into a braided area about 1,500 ft wide containing about one-third of the mainstem channels, then by 1996, the entire flow of the Matanuska River had swung 55 degrees at the bedrock point and been captured into this new course. Braiding continued to develop, and by 2006, active channels spanned 5,000 ft of the former flood plain.

The cause of the shifting river course and the probability of its persistence in this position is a complex question not yet answered despite the many causal explanations offered. The extent of braid plain boundary erosion between 1917 and 1949, prior to the southerly swing, suggests that the changes were broad in scope and already in motion by 1949. Individual events, including construction of erosion control, the 1964 earthquake, and gravel mining, may have locally influenced channel movement, but it is unlikely that any one of them can be considered a cause of the major change. Dike construction along the right bank at BPM 2 to protect downstream railroad bridges effectively cut off northward-eroding mainstem channels by 1949, deflecting flow back to the 1916 position. Expansion of that dike resulted in an effective barrier across 20–30 percent of the active braid plain by 1962. This dike clearly served to locally deflect flow toward the south, but the capacity of this effect to propagate upstream several braid plain widths is likely limited by the relatively steep slope and high sediment load of the river. Tectonic subsidence from the magnitude 9.2 earthquake of 1964 created a regional swath of down-dropped land centered in Prince William Sound that gradually decreased toward the northwest to a few miles beyond the Matanuska River (Plafker, 1969). This created a southeasterly ground surface tilt that ranged from 1 to 2 ft of offset across the lower Matanuska River. No specific evidence of channel change from this tilt has been proposed, but it is possible this condition encouraged movement toward the south. The timeframe for tilting to be effective has been explored in flume experiments (Kim and others, 2010) that showed the effects of a sharp step in a braided river were quickly neutralized in an actively migrating river with high sediment transport. The small magnitude of the tilt relative to the typical relief across the braid plain and the swift reworking of channels in the Matanuska River suggest that the impact of the regional tilt probably had a limited timeframe. A final event, gravel mining near the Old Glenn Highway bridge, appears in the 1985 image and may have locally impacted channel position there, but the effects of this extraction, or its cessation, seem unlikely to have had the capacity to drive the major shift in course 4 mi farther downstream.

Erosion Control Features

Erosion control along the Matanuska River has included deliberate attempts, such as dikes and riprap (fig. 25), and indirect actions, such as placement of railroad or highway fill. All detectable engineered or substantial erosion control features present in 2006 are presented in appendix A (ErosionControlFeatures shapefile), and examples are shown in figure 26. Date of placement ranges from the early 1900s to the past several decades. Most erosion control features present in 2006 consisted of short lengths of rirapped bank or a series of short spur dikes and were installed in response to a threatened or eroding bank. Exceptions include a dike complex near BPM 2 associated with the former Alaska Railroad line to Chickaloon; bank armoring near the Old Glenn Highway between Maud Road and Ye Old River Road (fig. 18); and a levee 1 mi downstream adjacent to the Old Glenn Highway that targeted flooding rather than bank erosion. Responsible groups range from State and Borough entities to homeowners. Evaluation of the effectiveness or condition of these features was beyond the scope of this report.

Figure 25. Erosion control present along the Matanuska River, southcentral Alaska, that include: (A) Series of dikes near Circle View Subdivision; (B) detail of dike installed near Circle View subdivision in 2007; (C) riprap at King Mountain State Recreation Area; (D) riprap along Glenn Highway embankment near Chickaloon showing erosion; (E) rock placed along bank at home near Sutton; and (F) car bodies used as erosion control near Circle View subdivision.

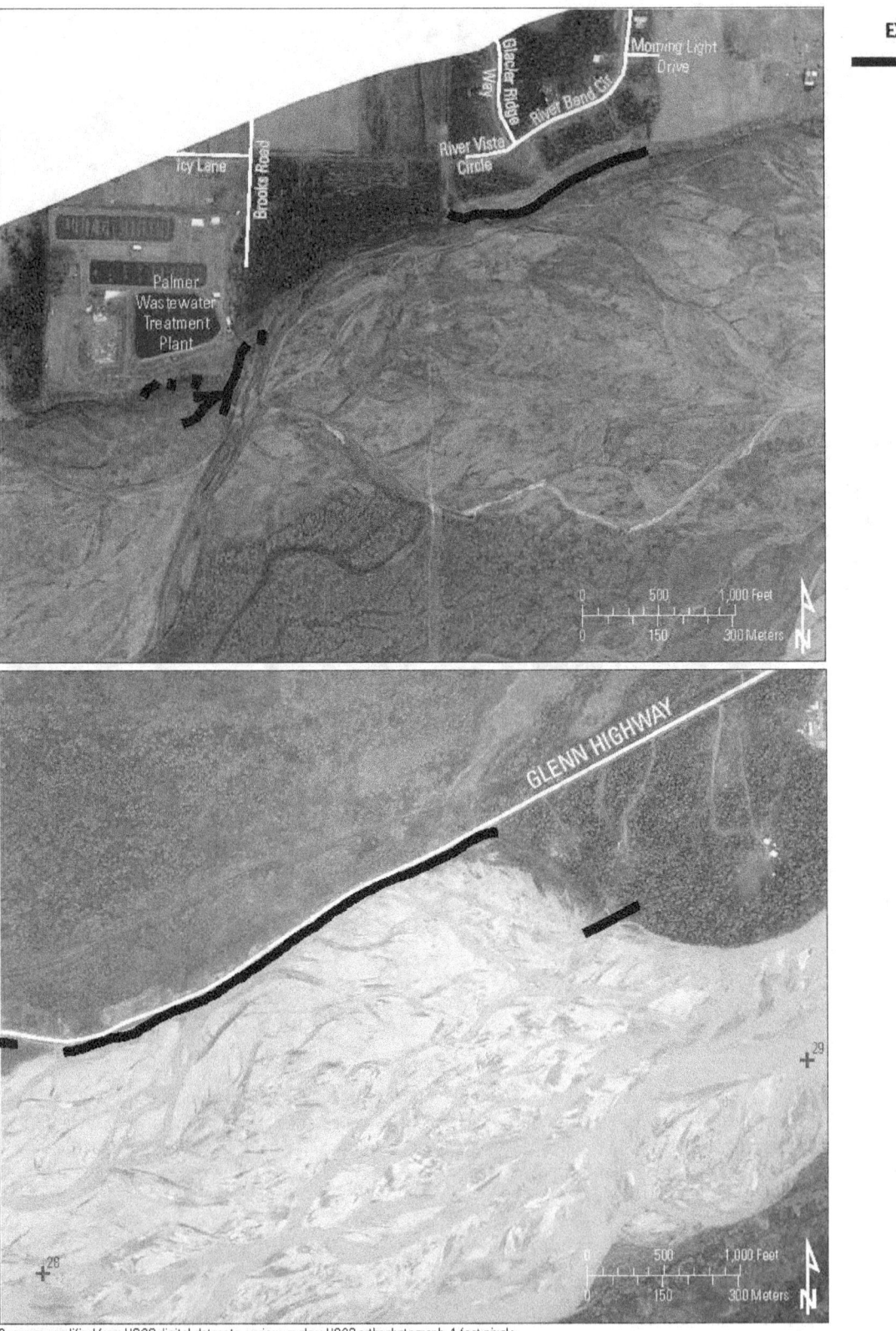

Figure 26.   Examples of mapped erosion control, Matanuska River, southcentral Alaska. (*A*) Dikes at a public facility and buried riprap at a residential subdivision. (*B*) Riprap along a Glenn Highway embankment and a dike provide erosion control near Glenn Highway Mile 69.

The inventory of 2006 conditions shows that the length of protected banks is relatively small, totaling 5.6 percent of the 138 mi of project area banks. Bank armoring was more extensive than dikes, consisting of 23 riprapped banks ranging in length from less than 100 ft to 0.7 mi for a total of 6 mi of bank protection. Dikes and levees were clustered into 11 groups of 1–7 structures, collectively totaling 2 mi in length. An additional 14 mi of bank is adjacent to the abandoned railroad bed, a portion of which is fill that contains coarse clasts likely to reduce erosion. Although not documented, the total length of informal bank protection (car bodies, rocks, tires, or placed trees) is estimated from field observations to be less than the length of the documented erosion control measures.

Erosion Hazards

Erosion has impacted or threatened a variety of residential, institutional, or historical structures (fig. 27). To provide a guiding objective for summarizing erosion hazards (the potential to erode), risks from bank erosion (the harm possible from erosion) were thought of in general categories of land use (undeveloped, residential, public facilities, or infrastructure) and geomorphic zones (river deposits and landslides) for this report. To preserve the ability for others to make different or more specific assessments of risk, if desired, this report discusses general patterns but presents additional detail in the banklines and geomorphic feature GIS files in appendix A. Although long reaches of bank are owned by agencies or Native groups, the significant amount of privately owned banks made retaining spatially detailed information more appropriate than composite assessment of longer reaches.

Predictive Variables

Assessment of erosion hazards requires an understanding of the driving factors of erosion and the resistance provided by the banks. Erosion can be driven by the capacity of the channel to erode and transport bank sediment, which is related to the depth and velocity of the river. Factors that affect bank resistance typically include bank material, bank grain size, moisture conditions, height and angle, vegetation, and landslide activity. In a wide braided river, a fundamental requirement for erosion is that the channel be against the bank. Factors that influence braid plain elevation, such as the presence of constrictions or aggradation/deposition within the braid plain, can have an effect on the tendency of the active channels to erode the bank rather than move within the braid plain.

Streamflow as a driver of Matanuska River bank erosion was examined in detail using instream measurements at a bank and in general using streamflow statistics. Detailed depth and velocity measurements taken during a period of active erosion near Circle View Subdivision (Conaway, 2008) noted velocities as high as 11 ft/s in a channel 5 ft deep at a visibly eroding bank on the outside of a channel bend, just downstream of its apex, and higher velocities and depths but no visible erosion at the apex. These measurements show the strength of hydraulic forces that overwhelmed the resistance of the 20 ft glacial terrace and effectively eroded and transported bank material during a period when streamflow was less than the 25th percentile of long-term values. Results such as this suggest that predicting future erosion using specific streamflow information or detailed estimates of hydraulic forces is not likely to be productive on a broad scale. The river is able to erode the banks at modest flows but does not always erode at high flows.

Matanuska River bank erosion was poorly correlated to measurable streamflow statistics, including peak streamflows and mean annual streamflows, over the subperiods between photographs. This prompted further consideration of the effects of floods. Inspection of a series of black and white, 1:12,000 scale aerial photographs taken within days after the large 1971 Granite Creek outburst flood showed little evidence of channel change or bank erosion. Although rapid erosion was visually observed during the highest flows of 2006 near Circle View Subdivision, smaller amounts of erosion also were observed for prolonged periods before and after the highest flows. These observations suggest that peak flooding is not the primary driver of erosion along Matanuska River banks.

Streamflow patterns for the river as a whole are well-characterized by gaging records, but frequent channel change can result in hydrologic variability not captured by measurements of the whole river. Streamflow in individual channels may increase disproportionately with the measured streamflow at the streamgage. A bankside channel may range from a minor slough to a main channel in a single season.

Given the capacity of the channel to erode, and poor performance of streamflow as an erosion predictor, a simple conceptual model for the Matanuska River expresses erosion as a factor of channel presence against the bank and the erodibility of the bank. Predicting the presence of the river against specific banks was limited by minimal historical data, but the braid plain age of 15 years provided a general sense of the frequency of channel movement within the braid plain. The lack of obvious patterns of channel occupation longitudinally along the river or over time did not suggest strong differences for specific locations. It is clear that decades without a potentially eroding channel against the bank are possible, but equally clear that the likelihood of a bankside channel is high within the lifespan of most structures.

Figure 27.   Structures impacted or threatened by bank erosion along the Matanuska River, southcentral Alaska, that include: (A and B) residential property and utility tower on eroding glacial terrace in Reach 1 near Circle View Subdivision, BPM 4.1; (C) damaged house on terrace in Reach 3 at BPM 24.0; (D) sandbags placed as temporary erosion control at eroded section of Glenn Highway in Reach 3 at BPM 23.5; and (E) historical cabin on presently stable bank at BPM 26.3 near Kings River shows extent of previous erosion.

Bank erodibility, expressed as general categories combining the physical factors of bank height and material, and a geomorphic factor, included because of good correlation with historical erosion, became the most useful consideration for erosion hazards for the entire study area. Bank grain size does not vary significantly along the river, except at a few areas of locally high boulder concentration, and was not considered a significant factor for predicting hazards among unconsolidated banks. Among other potential predictive factors, visible bank moisture, consolidated banks, and landsliding were limited in extent.

The relative importance of erodibility factors for the Matanuska River that can be drawn from the mapping for this report can be shown in a qualitative way (fig. 28). Relative erodibility within a factor was ranked using basic principles for the physical factors of bank material and height. Materials were grouped into categories termed unconsolidated sediment, less-resistant bedrock (which included non-bedrock features that provided significant erosion resistance), and bedrock on the basis of their erodibility. Geomorphic features were ranked on the basis of results of historical erosion, which showed more erosion at modern fluvial deposits than uplands, and more erosion at terraces than fans. The obviously high erosion resistance of bedrock makes bank material the most important erodibility factor, but the importance of bank height relative to geomorphic feature is difficult to discern. All features within the braid plain are subject to clear riverine hazards including flooding, as well as erosion, and are removed from consideration of braid plain margin erosion.

This analysis assumes that river processes continue as they have for the historical period, on the order of 50 years for hydrologic and photographic information and a total of about 100 years when vegetative indicators are included. No obvious trends in hydrology or other relevant factors were observed that suggest that extrapolating present conditions would be inappropriate.

Erosion Hazard Categories

The information in this report can be applied to erosion hazard assessment in several ways. General hazard categories and their geographic locations can be identified using the banklines and geomorphic features. Patterns and amounts of historical erosion can also help refine estimates of the potential extent of erosion into the bank over a period of time.

Using data from the GeomorphicFeatures shapefile (for closed-off areas) and the Banklines_2006 shapefile (for banks not closed-off at the back into areas), several useful hazard areas can be defined. Detailed definition of some hazard areas may require additional information or arbitrary choices from the user. Some of these hazard area categories are overlapping to accommodate a variety of possible applications. Specific combinations of erodibility factors (fig. 28) are interpretive,

such that the suite of categories discussed here for the anticipated need of river corridor management might be different from categories for management of specific reaches.

The most general hazard area that can be derived from the data is the corridor where bank erosion is possible, which can be defined by the combination of (1) the limits of bedrock mapped within the inner Matanuska River valley and (2) an arbitrary distance into upland features like glacial terraces or expansive fans where bedrock is not present. Beyond these limits, erosion by the river can be considered a negligible concern. This area can be mapped by selecting lines from the Banklines_2006 shapefile where the material is bedrock, the height is moderate or high, and the bedrock character is solid and by applying an arbitrary corridor width at other locations.

Within the erodible corridor, another type of hazard area consists of the braid plain, where erosion and flooding hazards are greatest. These areas can be selected from the GeomorphicFeatures shapefile where the feature type is braid plain. Selected historically flooded areas (feature name is historically flooded area), especially those surrounded by braid plain near the river mouth, can also be included in the braid plain, depending on user needs.

Adjacent to the braid plain, bank erosion hazard areas can be drawn using the erodibility factors relevant and mappable for the Matanuska River shown in figure 28. A simple analysis using an assignment of high, moderate, or low categorical values to the erodibility factors (material, height, and geomorphic feature), as well as to the values within each factor, results in an example ranking of the various combinations present (table 13). The greatest confidence can be placed in the respective ends of the spectrum of erodibility: unconsolidated, low banks that have historically flooded represent the highest erosion hazard, and all moderate or high solid bedrock banks represent a negligible erosion hazard. The latter distinction alone provides a strong constraint on the erodibility of the valley, because moderate or high solid bedrock banks flank 31 percent of the braid plain. However, sensitivity analysis using reversed factor weights for bank height and geomorphic feature shows that rankings of unconsolidated materials are sensitive to the factor weights. As an example, the difference in erodibility between a moderate-height terrace and a low tributary fan cannot be clearly defined from data collected for this report. For general planning purposes, a prudent assumption is that all unconsolidated material is easily erodible. For more detailed analysis of particular areas, the general principles shown in figure 28 can be used as a guide. Given two terraces of differing height, for example, the higher terrace is slightly less erodible in a statistical sense. A relatively simple grouping of the combination of factors mapped for the Matanuska River into bank erosion hazard categories of low, unconsolidated banks; moderate-high, unconsolidated banks; less-resistant bedrock banks, and bedrock banks is suggested by the distribution of features in table 13.

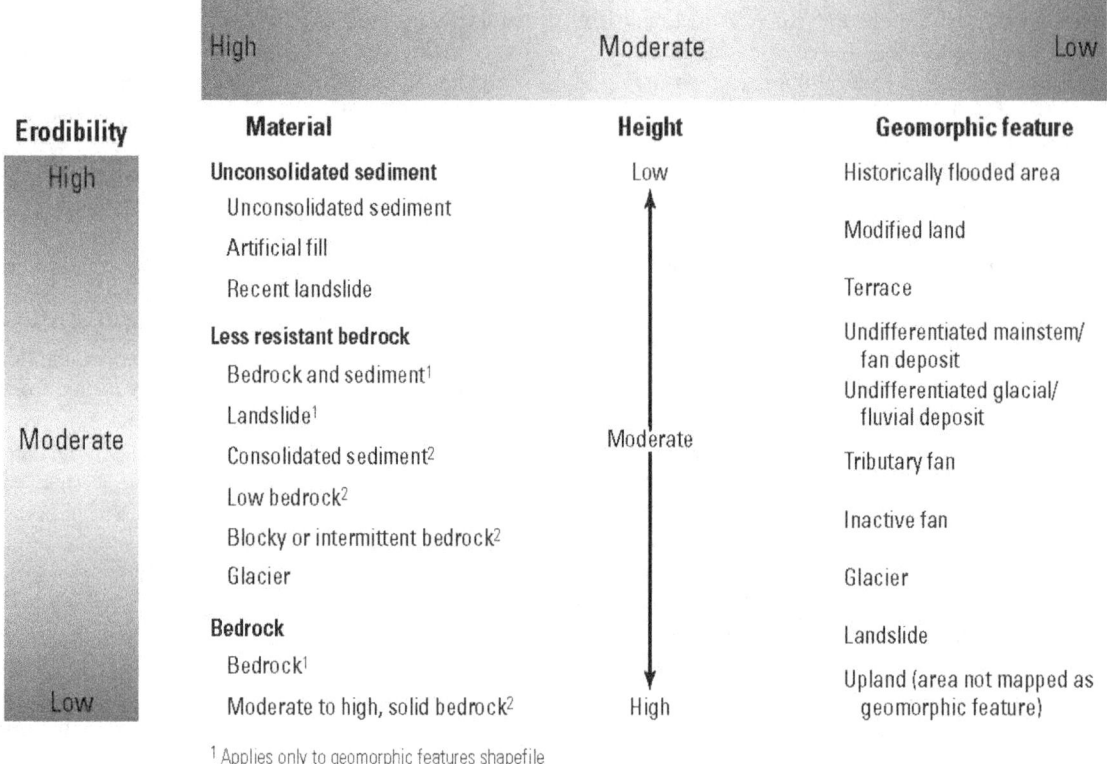

Note: Importance and erodibility values from high to low are relative rankings.

Primary bank erodibility factors

Factor importance

High   Moderate   Low

Erodibility

High

Moderate

Low

Material	Height	Geomorphic feature
Unconsolidated sediment	Low	Historically flooded area
Unconsolidated sediment		Modified land
Artificial fill		
Recent landslide		Terrace
Less resistant bedrock		Undifferentiated mainstem/ fan deposit
Bedrock and sediment[1]		Undifferentiated glacial/ fluvial deposit
Landslide[1]		
Consolidated sediment[2]	Moderate	Tributary fan
Low bedrock[2]		
Blocky or intermittent bedrock[2]		Inactive fan
Glacier		Glacier
Bedrock		Landslide
Bedrock[1]		Upland (area not mapped as geomorphic feature)
Moderate to high, solid bedrock[2]	High	

[1] Applies only to geomorphic features shapefile
[2] Applies only to banklines shapefile

Other erodibility factors (importance not ranked)

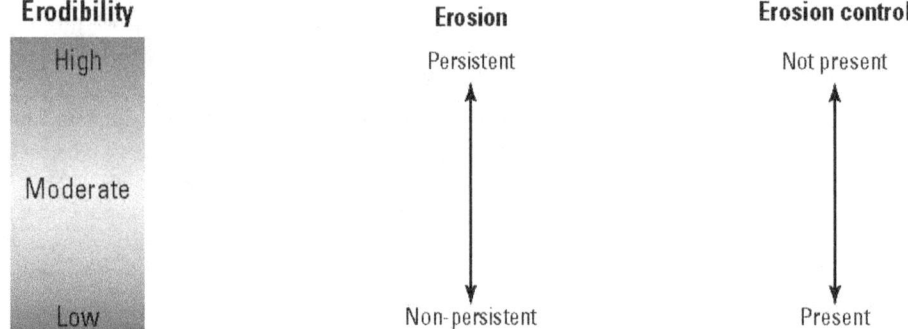

Erodibility

High

Moderate

Low

Erosion

Persistent

Non-persistent

Erosion control

Not present

Present

Figure 28.   Ranking of variables and characteristics associated with bank erosion along the Matanuska River, southcentral Alaska.

Table 13. Example bank erosion hazard categories for the Matanuska River, southcentral Alaska.

[Definitions for combinations of material categories in shapefiles are shown in figure 28. **Abbreviations:** –, bankline only, not mapped in geomorphic feature shapefile]

Material	Height	Geomorphic feature	Bank erosion hazard category
Unconsolidated	Low	Historically flooded area	Low, unconsolidated bank
Artificial fill or unconsolidated sediment	Low	Roadway, railroad, or modified	Low, unconsolidated bank
Unconsolidated	Low	Terrace	Low, unconsolidated bank
Unconsolidated	Low	Tributary valley	Low, unconsolidated bank
Unconsolidated	Low	Undifferentiated main stem/fan	Low, unconsolidated bank
Unconsolidated	Low	Undifferentiated glacial/fluvial	Low, unconsolidated bank
Unconsolidated	Low	Tributary fan	Low, unconsolidated bank
Unconsolidated	Low	Inactive fan	Low, unconsolidated bank
Unconsolidated	Low	–	Low, unconsolidated bank
Landslide	Low	Recent landsliide	Low, unconsolidated bank
Artificial fill	Moderate	Modified	Moderate-high, unconsolidated bank
Unconsolidated	Moderate	Terrace	Moderate-high, unconsolidated bank
Unconsolidated	Moderate	Tributary valley	Moderate-high, unconsolidated bank
Unconsolidated	Moderate	Undifferentiated main stem/fan	Moderate-high, unconsolidated bank
Unconsolidated	Moderate	Undifferentiated glacial/fluvial	Moderate-high, unconsolidated bank
Unconsolidated	Moderate	Tributary fan	Moderate-high, unconsolidated bank
Unconsolidated	Moderate	–	Moderate-high, unconsolidated bank
Unconsolidated	High	Terrace	Moderate-high, unconsolidated bank
Unconsolidated	High	Undifferentiated glacial/fluvial	Moderate-high, unconsolidated bank
Unconsolidated	High	Tributary fan	Moderate-high, unconsolidated bank
Unconsolidated	High	–	Moderate-high, unconsolidated bank
Artificial fill	High	–	Moderate-high, unconsolidated bank
Bedrock and sediment	Low	Tributary valley	Less-resistant bedrock bank
Bedrock and sediment	Low	Terrace	Less-resistant bedrock bank
Less resistant bedrock	Low	–	Less-resistant bedrock bank
Bedrock and sediment	Moderate	Terrace	Less-resistant bedrock bank
Landslide	Moderate	Landslide	Less-resistant bedrock bank
Less resistant bedrock	Moderate	–	Less-resistant bedrock bank
Landslide	High	Landslide	Less-resistant bedrock bank
Glacier	High	Glacier	Less-resistant bedrock bank
Consolidated	High	–	Less-resistant bedrock bank
Less resistant bedrock	High	–	Less-resistant bedrock bank
Bedrock	Moderate	Terrace	Bedrock bank
Bedrock	Moderate	–	Bedrock bank
Bedrock	High	–	Bedrock bank

To further refine the unconsolidated bank categories, banks in historically persistent erosion hotspots, such as near Circle View Subdivision, regardless of bank height, might be considered a higher hazard. It is unclear whether non-persistent hotspots have a higher probability of future erosion; these require individual assessment because some have been protected by erosion control, and some have fully eroded away. The high banks with variable presence of bedrock at their base, such as those on the right bank upstream of Palmer, can be considered a greater hazard than bedrock banks.

An example of how these hazard categories can be expressed in map form is shown in figure 29, where erosion hazard categories are shown at geomorphic features and along upland braid plain margins. The bank erosion hazard area is drawn to encompass braid plain bars and islands as well as the active channel belt, and bedrock uplands at the back of geomorphic features are shown to indicate where the extent of the corridor of possible erosion is clearly defined.

Additional hazards incidentally considered for this report are alluvial fan hazards and landslide hazards. Areas subject to these hazards can be can be depicted by selecting the areas with a feature name of tributary fan or landslide in the GeomorphicFeatures shapefile, respectively. Hazards at alluvial fans stem from the mobility of the stream across the fan, a general sense of which can be determined for Matanuska River tributaries from the various Feature ID categories mapped for different fan surfaces in the GeomorphicFeatures shapefile. Landslide mapping for this report was limited to published landslides adjacent to the river and opportunistic observations of recent movement, and should not be considered comprehensive.

Bank erosion hazards can be considered to extend the full depth of the geomorphic feature in areas where the river is confined in a canyon, but require an arbitrary depth estimate in unconfined areas. This study assumes that the erodibility of a terrace, for example, is consistent throughout the entire depth of the terrace because the materials and height of the terrace do not change. Other factors, such as the presence of erosion control, or major roadways like the Glenn Highway likely to be protected, may realistically limit the depth of erosion hazards for planning purposes. For unconfined areas, arbitrary depth estimates need to balance the unlimited erodibility of the banks with practical timespans. This estimate can consider average annual erosion rates (for example, see U.S. Department of Agriculture, Natural Resources Conservation Service [2004]), but it also needs to consider that the erosion has generally occurred as a swift, episodic process rather than a slow, progressive process.

The amount of erosion possible along banks can be estimated using the patterns and amounts of historical erosion of a similar erosion hazard category. The strongly localized nature of historical erosion along the Matanuska River suggests that average annual rates for a reach are likely to underestimate the potential erosion at a specific location. A model that emerges from the patterns of severe erosion along the Matanuska River is episodic, localized erosion of several hundred feet that can occur on an annual to decadal time scale. For example, at terraces near BPM 20 and 29, the river eroded more than 200 ft at part of the terrace and not at all elsewhere on the terrace. The erosion occurred within the 1 to 4 decades of the respective orthophotograph subperiods, but observations of fluvial process suggest that it could have occurred over a period of years rather than decades. Given the similarity in materials within a terrace and the lack of other predictor variables, the potential for erosion anywhere on those or similar terraces can be considered hundreds of feet within several years to a decade.

Erosion statistics computed for this report provide estimates of the extent of erosion concerns. Erosion detectable with aerial photographs (61 ft or more), occurred at 23 percent (31 mi) of the banks between 1949 and 2006, providing an estimate of the length of bank that might be affected by erosion in any given 50-year period. An analysis of the sensitivity of banks to erosion (for example, developed or undeveloped) would be required to estimate the length of banks that would require protecting in that period. A very coarse estimate of this can be obtained by the length of erosion control features installed by the time of this study. More than 8 mi of bank were sufficiently sensitive to erosion to require erosion control by formal or informal means. Persistent erosion was identified at selected locations, but at many others the river eroded enough to trigger action (for example, construction of dikes) and then moved away. This suggests that the persistently, severely eroded areas, such as along the left bank near the Old Glenn Highway and near Circle View Subdivision, can be identified as areas likely to require attention, but that other areas can appear suddenly.

Although this analysis has primarily focused on erosion detectable by comparing historical and recent orthophotographs, smaller amounts of erosion were observed to have a significant impact to private property and public infrastructure. Although the erosion between Kings River and Granite Creek (fig. 17) fell below the detection limit for mild erosion in the historical orthophotograph analysis, many structures in this reach were constructed on a terrace within 50 ft of the braid plain margin. Between 2006 and 2010, the erosion, although minor relative to overall historical Matanuska River erosion, resulted in the abandonment of two homes (fig. 27), imminent threat to many others, and emergency repair to a section of the Glenn Highway (fig. 27).

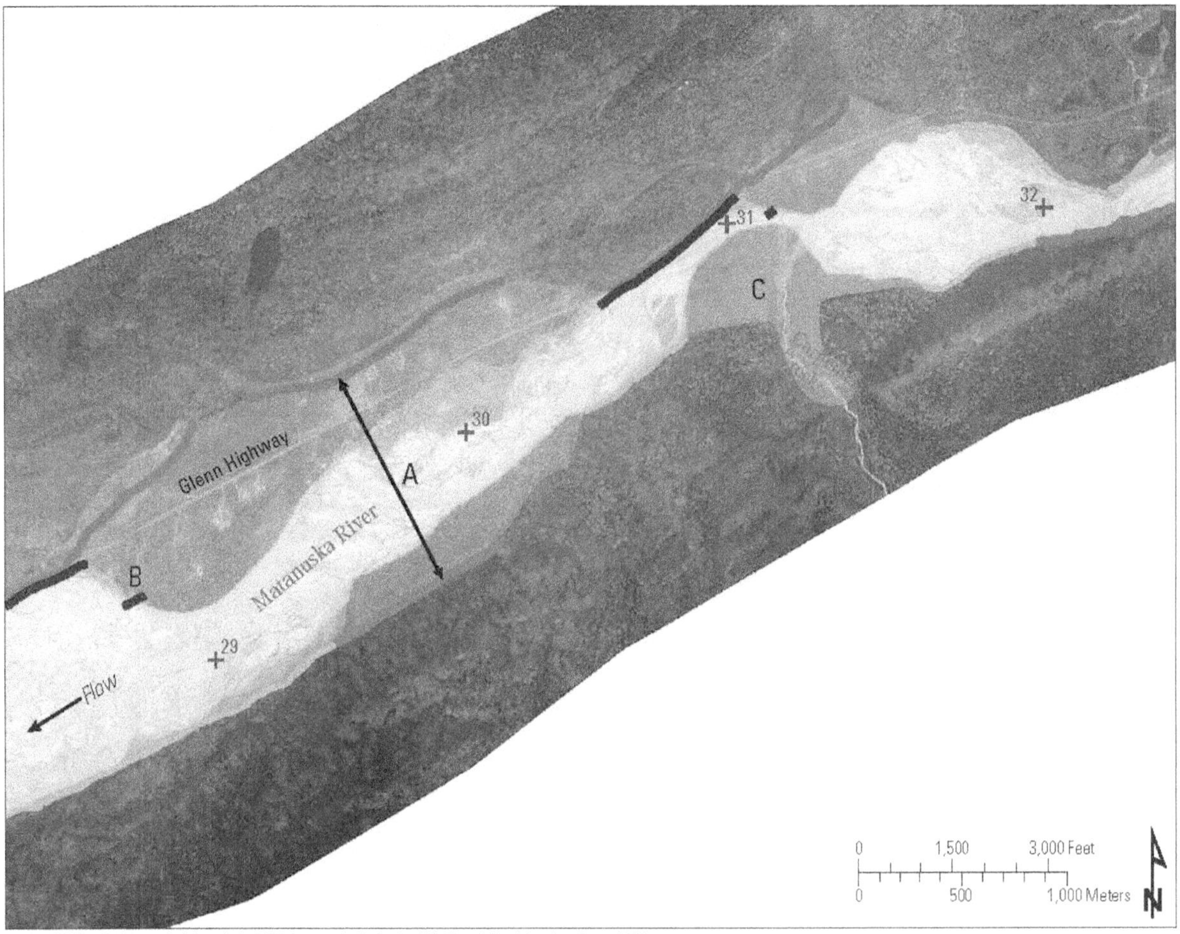

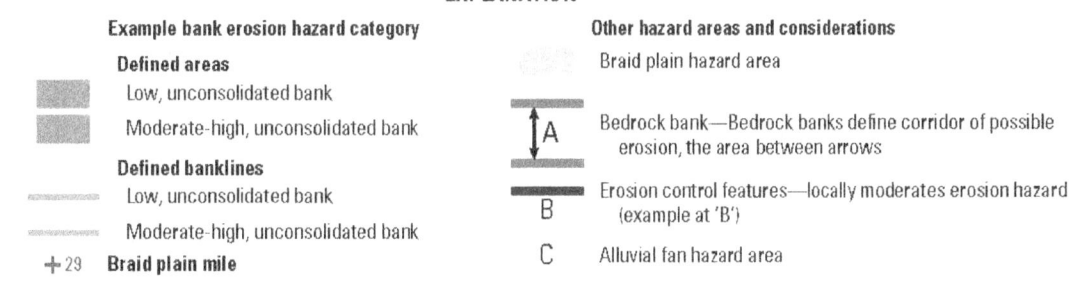

Imagery: USGS orthophotograph, 1-foot pixels, 2006 photography. Projection: UTM Zone 6, North American Datum 1983

EXPLANATION

Example bank erosion hazard category

Defined areas

Low, unconsolidated bank

Moderate-high, unconsolidated bank

Defined banklines

Low, unconsolidated bank

Moderate-high, unconsolidated bank

+ 29   **Braid plain mile**

Other hazard areas and considerations

Braid plain hazard area

Bedrock bank—Bedrock banks define corridor of possible erosion, the area between arrows

Erosion control features—locally moderates erosion hazard (example at 'B')

C   Alluvial fan hazard area

Figure 29.   Example erosion hazard map for the Matanuska River valley near Carpenter Creek, southcentral Alaska.

Summary and Conclusions

Banklines and geomorphic features mapped from 1949, 1962, and 2006 orthophotography and field estimates of bank height and material were used to generate maps of the braid plain and banks of the Matanuska River from the Matanuska Glacier to the river mouth. Bank erosion from 1949 to 2006 was computed from the banklines. An assessment of river processes guided interpretation of erosion hazards from the map products and erosion measurements.

Over the 57 years from 1949 to 2006, the actively braiding Matanuska River expanded its braid plain through bank erosion that was episodic and localized. Severe erosion occurred at lengths of bank an average of 0.5 mile long and ranged from 225 to 1,043 feet. Collectively, these erosion hotspots accounted for 64 percent of the erosion but only occurred at 8 percent of the banks. A major channel shift in the lower 3 miles of the river resulted in a 55 degree swing toward the south into a low-lying area and was not considered part of the erosion analysis area.

A high disturbance regime was typical within the braid plain, which had an average age of 15 years on the basis of vegetation ages present in 2006. Individual channels could span the entire width of the braid plain or be concentrated on one side, such that one or both banks could potentially be eroded at a time. Changes to the channel pattern occurred on an annual to decadal scale. The river reoccupied abandoned areas of the braid plain swiftly, regardless of forested cover. Planform changes from a braided to a more channelized river were observed near the Matanuska Glacier and may represent the local influence of changes to sediment or water discharge from the glacier.

The most common banks were unconsolidated sediments in modern river terraces and tributary fans along the confined parts of the Matanuska Valley, and glacial terraces in the lowland area of the valley. Bedrock banks greater than 10 feet high lined 31 percent of the study area banks.

Erosion was not correlated to peak streamflows or mean annual streamflow. Streamflows with capacity to erode and transport bank sediment occur regularly during the prolonged summer high-flow season. The frequency at which a channel occupies a bank-side position is important for erosion, but difficult to predict except in a general sense. Because historical patterns of Matanuska River bank erosion were not clearly linked to particular locations or easily quantifiable hydrologic, geomorphic, or anthropogenic factors, bank erosion should be considered possible any open-water season when an active channel flows along an erodible bank. Thus, for the purposes of determining erosion hazards, bank erodibility, as determined by general categories of bank material and height, and geomorphic feature type were the most relevant factors.

The primary products of this report are shapefiles showing the distribution of bank materials and height along the river, and the distribution of geomorphic features such as fans and terraces likely to present erosion hazards. These lines and polygons can be overlain strategically using the relative importance of the erodibility factors to develop erosion management categories depending on the degree of risk acceptable. Clear riverine hazards are presented by the braid plain, and clear erosion protection is provided by moderate and high bedrock banks. Additional hazards are presented by active tributary fans and active landslides. The detailed geographic data depicting the distribution of these various areas provide a scientific basis for erosion related management of the river corridor.

Acknowledgments

The authors thank the numerous individuals who contributed their knowledge of the river, granted access to their lands, and assisted in the field. In particular, the Matanuska-Susitna Borough and many private land owners graciously permitted access across their lands to the river. Kevin Sorensen of Glacier Properties provided river access and generously shared data and reports. Chuck Spaulding of NOVA shared river observations and rafting guidance. Brian Winnestaffer of Chickaloon Village Traditional Council cheerfully provided field assistance. Discussions and field trips with numerous individuals helped refine geologic and glacial geomorphology concepts, in particular Sara Kopczynski (Lehigh University), Dan Lawson (U.S. Army Corps of Engineers-Cold Regions Research and Engineering Laboratory), Mark Clark (U.S. Department of Agriculture, Natural Resources Conservation Service), Ellen Wohl (Colorado State University), and Jim O'Connor (U.S. Geological Survey). Dozens of participants of annual floating field trips helped refine concepts and shape presentation of the data.

References Cited

Alaska Department of Commerce, Community, and Economic Development, 2005, Alaska community database community summaries (CIS): Alaska Department of Commerce, accessed July 13, 2011, at http://www.commerce.state.ak.us/dca/commdb/CIS.cfm.

Anderson, J.L., and Bromaghin, J.F., 2009, Estimating the spawning distribution of Pacific salmon in the Matanuska River watershed, southcentral Alaska, 2008: U.S. Fish and Wildlife Service, Alaska Fisheries Data Series Number 2009–12, 43 p.

Barnes, F.F., 1962, Geologic map of the Lower Matanuska Valley, Alaska: U.S. Geological Survey Miscellaneous Geological Investigations Map I-359.

Clark, M.H., 2006, Soil survey of Chickaloon Village Lands Area, Alaska: U.S. Department of Agriculture Natural Resources Conservation Service, 137 p., and geospatial data.

Clark, M.H., and Kautz, D.R., 1998, Soil survey of Matanuska-Susitna Valley Area, Alaska: U.S. Department of Agriculture Natural Resources Conservation Service, 806 p. and geospatial data, text and geospatial data, accessed October 27, 2011, at http://soildatamart.nrcs.usda.gov/Manuscripts/AK600/0/MatanuskaSusitna.pdf and http://soildatamart.nrcs.usda.gov/, respectively.

Conaway, J.S., 2008, Bathymetric and hydraulic survey of the Matanuska River near Circle View Estates, Alaska: U.S. Geological Survey Open-File Report 2008-1359, 20 p. (Also available at http://pubs.usgs.gov/of/2008/1359/.)

Curran, J.H., Meyer, D.F., and Tasker, G.D., 2003, Estimating the magnitude and frequency of peak streamflows for ungaged sites on streams in Alaska and conterminous basins in Canada: U.S. Geological Survey Water-Resources Investigations Report 034188, 101 p. (Also available at http://pubs.usgs.gov/wri/wri034188/.)

Curran, J.H., McTeague, M.L., Burril, S.E., and Zimmerman, C.E., 2011, Distribution, persistence, and hydrologic characteristics of salmon spawning habitats in clearwater side channels of the Matanuska River, southcentral Alaska: U.S. Geological Survey Scientific Investigations Report 2011-5102, 38 p. (Also available at http://pubs.usgs.gov/sir/2011/5102/.)

Denner, J.C., Lawson, D.E., Larson, G.J., Evenson, E.B., Alley, R.B., Strasser, J.C., and Kopczynski, S., 1999, Seasonal variability in hydrologic-system response to intense rain events, Matanuska Glacier, Alaska, U.S.A: Annals of Glaciology, v. 28, p. 267-271.

Detterman, R.L., Pflaker, G., Tysdal, R.G., and Hudson, T., 1976, Geology and surface features along part of the Talkeetna segment of the Castle Mountain–Caribou fault system, Alaska: U.S. Geological Survey Miscellaneous Field Studies Map MF-738, 1 map sheet, scale 1:63,360.

Egozi, R., and Ashmore, P., 2008, Defining and measuring braiding intensity: Earth Surface Processes and Landforms, v. 33, p. 2,121–2,138.

Fahnestock, R.K., and Bradley, W.C., 1973, Knik and Matanuska Rivers, Alaska—A contrast in braiding, in Morisawa, M., ed., Fluvial geomorphology: Geomorphology Symposia Series, 4th, Binghamton, New York, 1973, Proceedings, London, England, Allen and Unwin, p. 221–250.

Freethey, G.W., and Scully, D.R., 1980, Water resources of the Cook Inlet basin, Alaska: U.S. Geological Survey Hydrologic Investigations Atlas HA620, 4 map sheets, scale 1:1,000,000.

Hodgkins, G.A., 2009, Streamflow changes in Alaska between the cool phase (1947–1976) and the warm phase (1977–2006) of the Pacific Decadal Oscillation–The influence of glaciers: Water Resources Research, v. 45, no. 6, p. W06502.

Kim, W., Sheets, B.A., and Paola, C., 2010, Steering of experimental channels by lateral basin tilting: Basin Research, v. 22, p. 286–301.

Lamke, R.D., 1972, Floods of the summer of 1971 in South-Central Alaska: U.S. Geological Survey Open-File Report 72-215, 88 p.

Matanuska-Susitna Borough, 2010, Matanuska River Management Plan: Matanuska-Susitna Borough Ordinance Serial No. 10-089, 97 p.

McGee, D.L., 1974, Evacuation of unnamed glacial lake contributed to 1971 Matanuska Valley flood: Alaska Department of Natural Resources Division of Geological Survey Mines Bulletin, v. XXIII, no. 5, p. 6–8.

Muhs, D.R., McGeehin, J.P., Beann, J., and Fisher, E., 2004, Holocene loess deposition and soil formation as competing processes, Matanuska Valley, southern Alaska: Quaternary Research, v. 61, no. 3, p. 265–276.

Neal, E.G., Hood, E., and Smikrud, K., 2010, Contribution of glacier runoff to freshwater discharge into the Gulf of Alaska: Geophysical Research Letters, v. 37, no. 6, p. L06404.

Northwest Hydraulics Consultants, 2004, Geomorphological Investigation of the Matanuska River – Palmer, Alaska, in U.S. Department of Agriculture Natural Resources Conservation Service, Matanuska River Erosion Assessment: MWH, Design Study Report Final, v. 2, variously paged.

Pearce, J.T., Pazzaglia, F.J., Evenson, E.B., Lawson, D.E., Alley, R.B., Germanoski, D., and Denner, J.D., 2003, Bedload component of glacially discharged sediment—Insights from the Matanuska Glacier, Alaska: Geology, v. 31, no. 1, p. 7–10.

Plafker, G., 1969, Tectonics of the March 27, 1964, Alaska earthquake: U.S. Geological Survey Professional Paper 543-I, 74 p.

Reger, R.D., and Updike, R.G., 1983, Upper Cook Inlet region and the Matanuska Valley, in Pewe, T.L., and Reger, R.D., eds., Guidebook to permafrost and quaternary geology along the Richardson and Glenn Highways between Fairbanks and Anchorage, Alaska: Fairbanks, Alaska, Alaska Division of Geological and Geophysical Surveys, Guidebook 1, p. 185–259.

Reger, R.D., Pinney, D.S., Burke, R.M., and Wiltse, M.A., 1996, Catalog and initial analyses of geologic data related to Middle to Late Quaternary deposits, Cook Inlet region, Alaska: Alaska Division of Geological and Geophysical Surveys Report of Investigation 95-6.

Restoration Science and Engineering, 2006, Hydrologic reconnaissance survey of tributaries of the Matanuska River: Palmer Soil and Water Conservation District, 50 p.

Thieler, E.R., Himmelstoss, E.A., Zichichi, J.L., and Ergul, Ayhan, 2009, Digital Shoreline Analysis System (DSAS) version 4.0—An ArcGIS extension for calculating shoreline change: U.S. Geological Survey Open-File Report 2008-1278, accessed July 13, 2011, at http://pubs.usgs.gov/of/2008/1278/.

Trainer, F.W., 1960, Geology and ground-water resources of the Matanuska Valley agricultural area, Alaska: U.S. Geological Survey Water-Supply Paper 1494, 116 p., 3 pls.

Trainer, F.W., 1961, Eolian deposits of the Matanuska Valley Agricultural Area, Alaska: U.S. Geological Survey Bulletin 1121-C, 34 p.

U.S. Army Corps of Engineers, 2003, Expedited reconaissance study, section 905(b) (WRDA 1986) Analysis, Matanuska River Erosion, Matanuska-Susitna Borough, Alaska, 30 p.

U.S. Census Bureau, 2009, 2009 Population estimates: U.S. Census Bureau, accessed February 24, 2011, at http://www.census.gov.

U.S. Department of Agriculture Natural Resources Conservation Service, 2004, Matanuska River Erosion Assessment: MWH, Design Study Report Final, v. 1 and 2, variously paged.

U.S. Department of Agriculture Natural Resources Conservation Service, 2004, Matanuska-Susitna Borough: U.S. Department of Agriculture Natural Resources Conservation Service, 1-meter pixel orthophotography.

U.S. Geological Survey, 2009, Water-resources data for the United States, water year 2009: U.S. Geological Survey, accessed July 13, 2011 at http://wdr.water.usgs.gov/.

U.S. Surveyor General's Office, 1915, Township No. 17 North, Range No. 1 East of the Seward Meridian, Alaska: U.S. Surveyor General's Office, 1 map sheet, scale 40 chains to an inch.

U.S. Surveyor General's Office, 1916, Township No. 16 North, Range No. 1 East of the Seward Meridian, Alaska; Township No. 17 North, Range No. 2 East of the Seward Meridian, Alaska; Township No. 18 North, Range No. 2 East of the Seward Meridian, Alaska: U.S. Surveyor General's Office, 1 map sheet each, scale 40 chains to an inch.

Western Region Climate Center, 2000, PRISM precipitation maps–1961–90, Mean Annual Precipitation, Alaska-Yukon: Western Regional Climate Center, accessed July 13, 2011 at http://www.wrcc.dri.edu/pcpn/ak.gif.

Western Region Climate Center, 2010, Western U.S. climate historical summaries: Western Regional Climate Center, accessed February 28, 2011, at http://www.wrcc.dri.edu/summary/Climsmak.html.

Wiedmer, M., Montgomery, D.R., Gillespie, A.R., and Greenberg, H., 2010, Late Quaternary megafloods from Glacial Lake Atna, Southcentral Alaska, U.S.A.: Quaternary Research, v. 73, p. 413-424.

Wiley, J.B., and Curran, J.H., 2003, Estimating annual high-flow statistics and monthly and seasonal low-flow statistics for ungaged sites on streams in Alaska and conterminous basins in Canada: U.S. Geological Survey Water-Resources Investigations Report 03-4114, 61 p. (Also available at http://pubs.usgs.gov/wri/wri034114/.)

Wilson, F.H., Hults, C.P., Schmoll, H.R., Haeussler, P.J., Schmidt, J.M., Yehle, L.A., and Labay, K.A., compilers; digital files prepared by Wilson, F.H., Hults, C.P., Labay, K.A., and Shew, N., 2009, Preliminary geologic map of the Cook Inlet region, Alaska–including parts of the Talkeetna, Talkeetna Mountains, Tyonek, Anchorage, Lake Clark, Kenai, Seward, Iliamna, Seldovia, Mount Katmai, and Afognak 1:250,000-scale quadrangles: U.S. Geological Survey Open-File Report 2009-1108, scale 1:250,000. (Also available at http://pubs.usgs.gov/of/2009/1108/).

Winkler, G.R., 1992, Geologic Map and Summary Geochronology of the Anchorage 1° × 3° Quadrangle, Southern Alaska: U.S. Geological Survey Miscellaneous Investigations Series Map I-2283, 1 map sheet, scale 1:250,000.

Appendix A. Names and Descriptions of GIS Files Published with this Report

[Files are available online at http://pubs.usgs.gov/sir/2011/5214]

File name	Short name	Description
RiverCenterline	River centerline	Centerline of the largest wetted channel of the Matanuska River in 2006 within the study area
River_MI	River miles	Points designating the location of even-mile-increments along the 2006 river centerline within the study area.
River_KM	River kilometers	Points designating the location of even-kilometer-increments along the 2006 river centerline within the study area.
BraidPlainCenterline	Braid plain centerline	Centerline of the 2006 Matanuska River braid plain within the study area.
BraidPlain_MI	Braid plain miles	Points designating the location of even-mile-increments along the 2006 braid plain centerline within the study area.
BraidPlain_KM	Braid plain kilometers	Points designating the location of even-kilometer-increments along the 2006 braid plain centerline within the study area.
Banklines_2006	2006 banklines	Lines designating the toe of slopes for banks and significant geomorphic features along the Matanuska River corridor within the study area.
BraidPlainMargin_1949	1949 braid plain margin	Line designating the position of the braid plain margin in 1949.
BraidPlainMargin_1962	1962 braid plain margin	Line designating the position of the braid plain margin in 1962.
BraidPlainMargin_2006	2006 braid plain margin	Line designating the position of the braid plain margin in 2006.
ErosionControlFeatures	Erosion control features	
Erosion	Erosion 1949–2006	Amount of braid plain margin change 1949-2006 computed from transects and projected onto 2006 braid plain margin position
GeomorphicFeatures	Geomorphic features	Polygons depicting mapped geomorphic features within the Matanuska River valley.
Uplands	Uplands	Single polygon enclosing the part of the study area that was not mapped. This area is generally outside the bounds of alluvial features associated with the Matanuska River and its tributaries.
GeomorphicReaches	Geomorphic reaches	Polygons outlining geomorphic reaches designated in this report for the Matanuska River.
ProjectBoundary	Study area boundary	Polygon outlining the study area for this report.
Railroad	Abandoned railroad	Lines designating the location of remnants of the abandoned railroad near or along the Matanuska River banks.